Soil Fertility Management Through Bioresource Technology

NIPA® GENX ELECTRONIC RESOURCES & SOLUTIONS P. LTD.
New Delhi-110 034

About the Authors

Dr. R. Singaravel, M.Sc.(Ag.), Ph.D., Post Doc. (Aus.), Professor of Soil Science and Agricultural Chemistry, Annamalai University Graduated during 1988 and completed M.Sc.(Ag.) during 1990 from the Faculty of Agriculture, Annamalai University. He was awarded Ph.D. in Soil Science and Agricultural chemistry from the Tamil Nadu Agricultural University during 1999. He did his Post Doc. at Murdoch University, Australia during 2010 under Endeavour Research Fellowship awarded by the Govt. of Australia. He was awarded with the Indian Society of Soil Science Award for the best Ph.D. during 1999. He has 29 years of research and teaching experience. He has published about 52 research articles in International Journals and 34 articles in National Journals. Visited countries like USA, Australia, China, Thailand and Sri Lanka for the presentation of research papers at International conferences. He has participated and presented 97 papers in National and International conferences. He has guided 8 Post Graduate and 4 Ph.D. scholars. He has prepared 5 research manual for students use at University level and published 6 books. He has operated several major research projects funded by UGC, DST, NMPB, TN State Govt. and other Private Agencies. He served as Joint Controller of Examinations, Annamalai University. At present he is serving as Director of Distance Education, Annamalai University.

Dr. V. Prasath, graduated from College of Agriculture, Pune (Mahathma Phule Krishi Vidyapeeth, Rahuri. Completed M.Sc.(Ag.) during 2000 and obtained his Ph.D. in Soil Science and Agricultural Chemistry during 2007 from Annamalai University. Worked as Project Fellow in UGC sponsored Research Project during 2002 to 2005 at Faculty of Agriculture, Annamalai University. He has published 4 research papers in International Journals and 8 research papers in National Journals. He has presented 15 research papers at National and International Conferences. At present working as Deputy Manager (Plantation) in Tamilnadu Newsprint and Papers Limited, Kagithapuram. Tamil Nadu.

Soil Fertility Management Through Bioresource Technology

R. Singaravel
V. Prasath

NIPA® GENX ELECTRONIC RESOURCES & SOLUTIONS P. LTD.
New Delhi-110 034

NIPA® GENX ELECTRONIC
RESOURCES & SOLUTIONS P. LTD.

101,103, Vikas Surya Plaza, CU Block
L.S.C. Market, Pitam Pura, New Delhi-110 034
Ph : +91 11 27341616, 27341717, 27341718
E-mail: newindiapublishingagency@gmail.com
Website: www.nipabooks.com

For customer assistance, please contact
Phone: + 91-11-27 34 17 17
Fax: + 91-11-27 34 16 16
E-Mail: feedbacks@nipabooks.com

ISBN: 978-93-95763-08-0

Composed and Designed by NIPA®.

Preface

After the green revolution, India became self-sufficient in the production of food grains, but to meet the future need of food grains for the burgeoning population, extensive cropping was practiced which affected the soil health. Need for a second green revolution has been felt by Scientists and policy makers in the recent years in India. To achieve this, it is essential to develop technologies having greater scope for improving soil fertility and productivity.

Maintenance of soil fertility and productivity is the dire need of present day farming as the level of soil organic matter is getting depleted due to over use of inorganic fertilizers, lesser use of organic manures, mono cropping, imbalanced fertilizer use etc. The area under multiple nutrients deficiencies are also increasing in recent years. As a result of degradation and depletion of soil fertility, agricultural production in many region is now failing to show positive response to increasing agricultural inputs. The Bioresource technology as a biosoftware can solve many constraints of soils and restore the fertility and productivity of crops. To renew the soil productivity, recycling of crop biomass and application of compost is a viable alternative. There is an absolute need to replenish soil by returning its lost share of organic matter which is recognized as a serious threat to sustainability. Huge amount of bio resources are available in India which could be effectively be recycled. Total amount of crop residue in India is estimated at 350 x 10^6 kg per year.

Bioresources contain the plant nutrient and the growth principles like enzymes and harmones, there is no such fertilizers which can supply all these together. Bioresources promote aggregation, permeability and water holding capacity of soil. The ion exchange and buffering capacity of soils is increased resulting in the nutrient holding capacity and stabilization of soil reaction. Soil enzymes and microbial population are enhanced which are important component in soils which as soil biocatalyst promotes the soil chemical and biochemical environment and nutrient availability to crop plants as well as other forms of life in soil whereby maintains soil fertility and sustainability. The organic acids released during decomposition chelate the metallic nutrient and make them easily available for absorption by the crops. Bio fertilizers supplements nutrients and help to check the nutrient deficiencies through nutrient fixation, solubilization of plant nutrients and stimulates plant growth by excreting growth harmones.

This book offer ten chapters which comprises potential benefits of various bioresources for the improvement of soil fertility in terms of physico-chemical properties, macro and micro nutrient availability, soil microbes and enzymes, crop growth, yield and plant nutrient uptake.

Efforts have been made to compile the information so as to make the book more useful. However perfection is never achieved and any suggestion for further improvement by the readers will be welcomed. The authors are also indebted to Authorities of Annamalai University for their support and encouragement.

Authors

Contents

1

Introduction

Around the world, nearly one billion people live along the 3,12,000 Km long coastline. The Indian coastline runs around 8,129 Km distributed along nine coastal states, two groups of islands and two union territories. The coastal ecosystem of India forms a very valuable resource community, supporting the livelihood security of several million rural poor and also contributing the national economy in a large measure. The latest estimates by National Bureau of Soil Survey and Land Use Planning (NBSS & LUP), Nagpur, India, using Global Assessment of Soil Degradation (GLASOD) guidelines indicate that the coastal area in the country is spread over an area of 11.8 million hectares (Mishra, 2005). Yadav *et al.* (1983) have reported about 3.1 M ha and 2.1 lakh ha of coastal salinity in India and Tamilnadu respectively.

Almost the entire coastal tract suffers from soil salinity, alkalinity, acidity, water logging, toxic element concentration, problems due to sand dunes and erosion, besides climatic adversities. In the coastal regions, the soils are rich in salts due to the presence of saline ground water table present at shallow depth and also due to the ingress of brackish seawater. Such salinity besides affecting the nutrient availability and physical conditions of the soil imposes a stress on growing crops that can lead to decreased yields and in severe cases there is a complete crop failure.

The high salt content and / or high pH cause changes in solubility, availability and efficiency of plant nutrients in coastal salt affected soils. Due to high pH and salinity, the rate of organic matter mineralization, urea hydrolysis and nitrification rates are poor owing to reduced bacterial activity. In addition, high N loss through volatilization of applied fertilizers and reduced uptake capacity of plant roots under stress condition leads to poor N use efficiencies in these soils (Sen and Bandyopadhyay, 1986).

Under high pH, the phosphorus present as PO_4^{3-} ions are known to be physiologically unavailable and may often be the limiting factor for plant growth. Further, the uptake of P by plant may also be severely affected due to the competitive inhibition caused by Cl^- ions for absorption sites. Prevalence of high pH, presence of $CaCO_3$ and low organic matter content decreases the availability of most of the micronutrient cations (Fe, Mn, Cu and Zn) as well. It has been observed that each unit increase in pH decreases the divalent cations by 100 times and trivalent cations by 1000 times (Lindsay, 1972). In almost all types of salt affected soils, there is a negative correlation between soil pH and available micronutrient cations.

In the coastal salt affected soils, the soil biology, including the population of microbes and enzymatic activities, which are known to play a greater role in nutrient availability and crop growth, are adversely affected largely due to the salinity on osmatic and ionic stress induced by the salts (Kumar and Kapoor, 1995; Garcia *et al.*, 2000).

It must not be lost sight of, at same time, the coastal areas have an important role to play in meeting out the demands of increased production for the burgeoning population. The coastal soils of India is probably degraded more than ever before due to indiscriminate and unscientific use of the lands for agricultural and its allied activities. The above fact suggests that the concerted efforts must be made on scientific nutrient management strategies for judicious exploitation of the degrading land resources of coastal ecosystem for sustainable productivity of crops.

The coastal areas of India are mainly monocropped growing rainfed rice during kharif (rainy) seasons. Under the prevailing soil and nutrient related constraints, the coastal farmers are exploiting these degraded lands with the unscientific management practices and achieving a low yield of around 2.1 t ha^{-1}. The estimated rice production target for coastal regions of India for 2010 AD is around 45 million tonnes and can be achieved only if the current productivity level of 2.1 t ha^{-1} is to be substantially increased within the short period of time (Yadav, 2004). This will also aid to meet India's projected rice production target of 140 million tonnes for 2025 AD, which can be achieved only by increasing the rice production by over 2.0 Mt per year in the coming decade. Under such situation these coastal salt affected soils offer greater avenue for the increasing rice production.

The coastal ecosystem is endowed with a variety of natural resources that can be effectively utilized for sustainable crop production in coastal salt affected soils. The role of organic matter to supplement amendments or as a source of nutrients and to increase the efficiencies of applied nutrients have been well documented by several workers (Yadav, 1980; Bandyopadhyay and Rao, 2001). Bio-resource technology as a biosoftware, can solve many constraints of soil by way of supplying the plant nutrients and the growth principles like enzymes and hormones, where there are no such fertilizers, which can supply all these together.

Bio-resource technology will have the following benefits in restoring the fertility of degraded coastal soils. This technology will promote the soil physical properties. The ion exchange and buffering capacity of the soil are increased resulting in the nutrient holding capacity and stabilization of soil reaction. Soil enzymes and microbial populations are enhanced which are important component in soils for maintaining soil fertility and sustainability (Burns, 1982). The organic acids released during decomposition chelate the metallic nutrient and make them easily available for absorption by the crops (Tan, 1986). Biofertilizers supplement nutrients and help to check the nutrient deficiencies through nutrient fixation, solubilization of plant nutrients and stimulates the plant growth by excreting growth hormones.

In restoring the soil fertility and productivity of coastal salt affected soils for sustainable crop production using bio-resources, the results of various Laboratory, pot and field experiments were discussed with supporting literatures that have accumulated in the past and were synthesided into ten chapters under different headings.

2

Effect of Bioresources on the Physical & Physico-chemical Properties of Soil

The coastal salt affected soils are low in productivity due to various soil fertility constraints that affect the crop growth and yield. The high salt content &/or high pH cause changes in solubility, availability and efficiency of plant nutrients, besides deteriorating the physical condition of soil. The soil biology (microbial population and enzymatic activities) also severely affected in these salt affected soils. The bio-resources are known to improve soil physical, chemical & biological properties of the soil. The maximization of rice production and considerable improvement in quality and composition of rice in coastal salt affected soil with the use of bio-resources have been reported by several research workers.

Soil Physical Properties

The importance of physical properties of soil on plant growth is widely recognized. The salinity and alkalinity of soil detoriates the physical condition of soil and as such affects the crop production in several ways. Poor physical properties, high pH, ESP and low organic matter are the main characteristics of coastal saline soils of India (Bandyopadhyay and Rao, 2001). It is well known that binding of soil particles into stable aggregates is essential for optimum soil tilth- a precursor to high productivity (Bellakki and Badanur, 1997; Celik *et al.*, 2004; Selvi *et al.*, 2005). Several studies carried out have shown the beneficial effects of addition of organic matter on physical properties of the soil.

Decrease in bulk density and increase in air porosity due to incorporation of crop residues, green manuring and application of farmyard manure was reported by More (1994)

Von Unexlkull and Mutert (1992) summarized the results of long term field experiments that besides supplying essential nutrients, organic manures improved the soil physical environment by way of i) improving soil structure, water holding capacity and soil aeration and ii) moderating soil surface temperature.Prasad (1994) reported that the bulk density decreased with the application of farmyard manure by improving the soil porosity.

Bellakki and Badanur (1997) studied the effect of organic and inorganic fertilizers on the physical properties of vertisol and reported that application of sunhemp or farmyard manure in combination with inorganic fertilizers decreased the bulk density of the soil.

Physical properties like bulk density, porosity, void ratio, permeability and hydraulic conductivity were significantly improved when FYM (10 t ha^{-1}) was applied in combination with chemical amendments, resulting in enhanced rice and wheat yields in sodic soil (Hussain et al. 2001)

Green manuring either with *Sesbania rostrata* or *Sesbania aculeata* improved the soil physical conditions by reducing bulk density and increasing soil aggregation and hydraulic conductivity (Mandal *et al.,* 2003). Celik *et al.* (2004) revealed that the application of farmyard manure @ 25 t ha^{-1} or compost made up of grasses, stubbles and plant leaves @ 25 t ha^{-1} significantly increased the porosity, soil aggregation and water retention capacity and decreased the bulk density.

Manna *et al.* (2005) reported that the application of 100% NPK through inorganic fertilizers + FYM @ 10 Mg ha^{-1} reduced the bulk density of the surface (0-15 cm) soil, which facilitated build up of soil organic matter as these two parameters are inversely proportional to each other.

Reduced bulk density and increased porosity, hydraulic conductivity, water holding capacity, aggregate stability and water stable aggregates due to the application of 100 % NPK + FYM @ 10 t ha^{-1} was reported by Selvi *et al.* (2005).

Bagayoko (2012) revealed that the use of organic manure, particularly poultry droppings and cow dung, helps to maintain the soil's physical state and is a major contributing factor to plant nutrients (Bagayoko, 2012). Tilahun Tadesse et al. (2013) revealed that that application of 15 t·FYM·ha"[1] significantly increased soil organic matter and available water holding capacity but decreased the soil bulk density, creating a good soil condition for enhanced growth of the rice crop. Wang *et al.* (2014) found that a mixture of organic wastes decreased bulk density by 11 % and increased the total porosity by 25 per cent.

Ondrasek and Rengel (2021) demonstrated that he use of organic amendments on saline soils improved porosity, hydraulic conductivity, soil aggregate stability and permeability.

Soil Physico-Chemical Properties, Cation Exchange Capacity and Exchangeable Sodium Percentage

Evaluation of pH, electrical conductivity, soil organic carbon and available nutrients of soil is deemed essential for assessing chemical aspects of soil quality, which are important as they provide measurement of the ability of soil to supply nutrients and to buffer against chemical additives of amendments (Gregorich *et al.*, 1994). High pH, EC and ESP adversely affects crop growth by reducing nutrient availability and decreasing the root volume of plants. Addition of organics for improving the physico- chemical properties and better crop production was well documented (Minhas *et al.*, 1995; Bandyopadhyay and Rao, 2001; Yadav, 2004).

Soil reaction

Ramaswamy (1981) indicated that composted coirpith as an amendment significantly reduced the physico chemical properties with reduction in pH, EC & ESP of the sodic soil. Savithri *et al.* (1991) reported that raw coirpith application lowered the soil pH due to its slightly acidic nature. A favorable reduction in soil pH of sodic soil with the application of *Sesbania aculeata* + rice husk or *Sesbania aculeata* + wheat straw was reported by Anand Swarup (1992).

More (1994) recorded reduction in pH due to addition of biogas slurry, press mud and wheat straw to a sodic soil. The maximum reduction in soil pH from 9.0 to 8.4 was recorded with the FYM and press mud application.

Srikanth *et al.* (2000) observed a significant reduction in pH of alfisol from 7.24 to 7.04 with FYM application (at the rate to supplement 50 % recommended phosphorus) due to the acidifying effect of organic acids produced during the course of decomposition. Yaduvanshi (2001) reported a significant decrease in soil pH of a sodic soil with the continuous application of green manures (*Sesbania aculeata* or FYM @ 10 t ha^{-1}) in a rice- wheat cropping system. Singaravel *et al.* (2001) reported a reduction in soil pH due to the application of *Casuarina* needles along with gypsum amendment in coastal saline soil. Wang and Yang (2003) reported a slight decline in pH due to the addition of 33% N + green manuring with *Sesbania rostrata* or clover.

Bhrat Singh and Pandey (2004) reported that application of gypsum and green manuring with dhaincha in a waterlogged sodic soil reduced the pH from 9.7 to 9.3. Rathore *et al.* (2004) revealed that the application of farmyard manure @ 10 t ha^{-1} resulted in reduced soil pH due to the organic acids release on its decomposition.

Singh et al. (2006) reported that the value of bulk density, pH and electrical conductivity decreased in all the integrated nutrient-management practices namely 25% RDN through pressmud + 75% RDF through inorganic fertilizers, and 50% RDN through pressmud + 50% RDN through inorganic fertilizers compared with its initial soil status.

Wang *et al.* (2014) reported that a mixture of organic wastes decreased the soil EC, and ESP by 87%, and 71%, respectively than the control. Ahmad et al.

(2015) reported that the application of gypsum and farmyard manure have improved soil physic chemical properties, including pH and EC.

Khatun et al (2019) reported that the released organic acids during decomposition of organic manure aid the mineralization of more organic mineral component of the soil. In a saline soil (10.6 dS m^{-1}) irrigated with a slightly saline water (4.28 dS m^{-1}), organic amendments, including vermicompost and cow dung, were found to improve the soil EC (3.37 dS m^{-1}) and pH, thus increasing maize growth compared to untreated control.

The integrated application of gypsum with compost has been reported to reduced EC, ESP, pH of saline sodic soil (Kitila et al., 2020). Rajendra Bhanwaria *et al.* (2022) evidenced that vermicompost @ 5 t ha^{-1} significantly decreased the bulk density and pH of soil.

Santosh Yadav *et al.* (2022) showed that the application of Zn enriched FYM decreased the pH and SAR significantly.

Electrical conductivity

Bandyopadhyay and Bandyopadhyay (1984) observed that regular application of FYM, green manures and crop residues to coastal saline soils resulted in significant reduction in electrical conductivity over the non amended soils. Saravanan *et al.* (1991) revealed that addition of composted coirpith caused reduction in pH and salt concentration (EC) and increased nutrient availability in the soil.

Polara *et al.* (2002) reported a significant reduction in EC of a coastal saline sodic soil due to the application of FYM @ 25 Mg ha^{-1} combined with gypsum @ 25 per cent gypsum reqirement. The reduction in EC accounted 71 per cent lesser than the control.

Singaravel and Prasath (2004) reported a significant reduction of EC from 1.90 to 1.22 dS m^{-1} in a coastal sodic soil due to the application of *Sesbania rostrata* and enriched farmyard manure as a substitution for 25 % of N and P for rice respectively. Kaushik *et al.* (2005) through an incubation experiment showed that application of rice husk or farmyard manure in the soil @ 2% W/W basis helped in lowering down the EC significantly from 4.0 to 3.3 dS m^{-1}.

Shaaban *et al.* (2013) demonstrated that gypsum application, along with humic acid and organic manure, in a saline-sodic soil has improved soil pH, SAR, EC (2.65 dS m^{-1} from 6.35 dS m^{-1}) and, consequently, the root growth and yield of rice.

Kabirinejad *et al.* (2014) reported that crop residue incorporation significantly decreased soil pH and EC. Adane *et al.* (2019) found that maximum reduction in soil pH (14.14%), ECe (48.99%) and ESP (16.20) was recorded in soils treated with combined 30 tons FYM ha^{-1} and gypsum (125% GR) treatments.

Bayoumy et al. (2019) proved that gypsum application, in conjunction with biochar and compost tea (containing organic matter and beneficial microbes), helped in ameliorating the effect of soil salinity on wheat production by reducing the EC, SAR and ESP.

Cation Exchange Capacity (CEC)

Organics by their high exchange properties are known to increase the CEC of the soil. Thompson *et al.* (1989) accounted for contribution of organic matter on CEC would be around 50 per cent. The important physico chemical properties *viz.* CEC and soil buffering capacity are influenced by the soil organic matter content (Palaniappan and Natarajan, 1993).

Clarson *et al.* (1984) revealed that application of composted coirpith increased the soil properties like CEC and available N, P and K in the soil. In a long term experiment, Bellakki *et al.* (1998) found that combined addition of various organics along with inorganic sources of nutrients resulted in significant increase in CEC of rice cultivated soil over control. The highest CEC was observed with addition of 75 percent N and remaining 25 per cent supplied through cowdung slurry.

Babhulkar *et al.* (2000) revealed that a marked increase in CEC of the soil by the application of higher rates of FYM. Neelam Sharad and Verma (2001) reported that addition of *Lantana camara*, an obnoxious weed as green manure @ 30 t ha^{-1} improved the CEC of the soils in rice- wheat cropping system.

Bulluck III *et al.* (2002) revealed that the CEC of the soil increased with the application of mixed yard waste- poultry litter compost, uncomposed cattle manure, mixed hay- cattle manure compost or composted gin trash in a sodic soil. Prakash *et al.* (2002) found that the organic manure applications, especially FYM registered relatively higher CEC as compared inorganic fertilizers.

Zhao *et al.* (2019) reported that the CEC and SOC increased by 8% and 22%, respectively in 0-20 cm soil layer by straw incorporation compared to control.

Exchangeable Sodium Percentage (ESP)

Salt toxicity is one of the major edaphic factors limiting crop production and eco- environmental quality in saline and / or sodic soil throughout the world (Liang *et al.*, 2005). The tissues of plants growing in saline media accumulate Na^{++} and Cl^{-} ions and inhibit the uptake of other essential plant nutrients. The application of organic matter in saline soils can accelerate Na^{++} leaching, decrease the exchangeable sodium percentage (ESP) and electrical conductivity and increase water infiltration, water holding capacity and aggregate stability (El-Shakweer *et al.*, 1998).

Selvi and Augustine Selvaseelan (1991) showed that application of composted coirpith considerably reduced the exchangeable sodium percentage of soil. Savithri *et al.* (1991) reported that application of composted coirpith @ 10 t ha^{-1} significantly reduced the ESP, besides improving the Ca and Mg status. Anand Swarup (1992) reported a significant reduction in the ESP of the soil and improvement in the exchangeable Ca^{++} and Mg^{++} with the addition of green manure (*Sesbania aculeata*) together with rice husk in a sodic soil.

Minhas *et al.* (1995) found a reduction in ESP and pH of an alkali soil with the application of gypsum @ 50% GR along with FYM @ 7.5 t ha^{-1}. The soluble Ca content increased significantly with the same amendments. Tripathi (1998) revealed that the integrated use of water hyacinth or pressmud or FYM along with chemical fertilizers significantly reduced the ESP and pH, besides saving 50 per cent of nitrogen fertilizers.

Bharambe *et al.* (2001) reported a considerable reduction in ESP of the soil after incorporation of green manure (dhaincha) @ 15 t ha^{-1}. Singaravel *et al.* (2001) reported a reduction in ESP of soil with the application of *Casuarina* needles along with gypsum.

Singaravel and Prasath (2004) reported a significant reduction of ESP in a coastal sodic soil due to the application of *Sesbania rostrata* and enriched farmyard manure as a substitution for 25 % of N and P for rice respectively. Sultan Singh *et al.* (2005) found that addition of FYM @ 15 t ha^{-1} and gypsum @ 100% GR in a light textured soil irrigated with sodic water significantly reduced the ESP.

Ahmad *et al.*(2015) showed that the application of gypsum and farmyard manure reduced the hydraulic conductivity and SAR, after 3 years of application on a saline-sodic soil field condition.

María Alcívar *et al.*(2018) showed that electrical conductivity (ECe), sodium adsorption ratio (SAR), and exchangeable sodium percentage (ESP) decreased significantly with gypsum (G), humic substances (HS) and biochar(B) application treatments.. When compared to the control, ESP decreased 11-fold in the G treatment, and 9–13-fold in the B + G; B + HS; and B + HS + G treatments.

Bayoumy et al. (2019) proved that gypsum application, in conjunction with biochar and compost tea (containing organic matter and beneficial microbes), helped in ameliorating the effect of soil salinity on wheat production by reducing the SAR, EC and ESP.

LABORATORY INCUBATION EXPERIMENTS

The coastal saline sodic soil exhibits low status of N and P due to the salinity and sodicity. The unfavorable soil conditions also reduces the efficiencies of applied nutrients. Bio-resources with their potentiality to improve the physical, chemical and biological properties of the soil will enhance the availability of plant nutrients. With the view to increase the availability and efficiencies of these nutrients, two separate incubation experiments one each on nitrogen and phosphorus were conducted using various bio-resources. For nitrogen incubation, the green manure *Sesbania rostrata*, composted coirpith and *Azospirillum* were used as bio-resources. For the phosphorus incubation, super phosphate enriched FYM, Casuarina needles and *Phosphobacterium* were used as the bio-resources.

Incubation Experiment I (Nitrogen)

In a Completely Randomised Design with three replication the following treatment viz., T_1-Absolute control, T_2-100% NPK, T_3- 75 % N + 25% N through *Sesbania rostrata* + PK, T_4- 75% N + *Azospirillum* + PK, T_5- 75% N + 25% N through composted coirpith + PK, T_6- 50% N + 50% N through *Sesbania rostrata* + PK, T_7- 50% N + *Azospirillum* + PK, T_8- 50% N + 50% N through composted coirpith + PK, T_9- 25% N + 75% N through *Sesbania rostrata* + PK, T_{10}-25% N + *Azospirillum* + PK, T_{11}- 25% N + 75% N through composted coirpith + PK were studied.

Incubation Experiment II (Phosphorus)

The P incubation experiment was conducted in a Completely Randomised Design with three replications. The following treatments viz., T_1. Absolute control, T_2. 100% NPK, T_3. 75% P + 25% P through enriched FYM + NK, T_4. 75% P + *Phosphobacterium* + NK, T_5. 75% P + 25% P through Casuarina needles + NK, T_6. 50% P + 50% P through enriched FYM + NK, T_7. 50% P + *Phosphobacterium* + NK, T_8. 50% P + 50% P through Casuarina needles + NK, T_9. 25% P + 75% P through enriched FYM + NK, T_{10}. 25% P + *Phosphobacterium* + NK, T_{11}. 25% P + 75% P through Casuarina needles + NK were studied.

The incubation experiment was carried out by taking 250 g of soil in an incubation bottles added with the treatments as listed above. Moisture was maintained at 60 per cent moisture holding capacity. Periodic soil samples at 15 days interval were taken and analysed for the properties like pH, EC, organic C, available major nutrients N,P, K and micro nutrients viz., Fe,Mn,Cu and Zn. The results emanated from the experiments are discussed with suitable literatures.

INCUBATION EXPERIMENTS

The experimental soil of the present study represented saline-sodic nature. It is obvious that improvement in the physico-chemical properties of these soils is the prerequisite for better crop nutrition. All the bio-resources used in the incubation experiments significantly improved the physico-chemical properties of the soil (Table 1 & 2).

In both the incubation experiments carried out for N and P nutrition, the various bio-resources application greatly reduced the soil reaction (pH), electrical conductivity (EC) and exchangeable sodium percentage (ESP) of the coastal salt affected soil. The initial soil had the pH of 8.97, EC of 1.94 dS $m^{-1}$and ESP of 31.44. Due to the bio-resources application, a reduction in soil pH (8.15 - 8.52), EC (1.22 - 1.49 dS m^{-1}) and ESP (17.95 - 25.56) in the nitrogen incubation experiment and in the phosphorus incubation experiment, a reduction in the range of 8.21 to 8.64 in pH, 1.28 to 1.74 dS m^{-1} in EC and 19.75 to 27.92 in ESP was recorded. The added green manure as the substitution for 25 per cent N in the N incubation experiment and enriched FYM as the substitution for 25 per cenet P in the P incubation experiment excelled over all other bio-resource treatments in reducing the pH, EC and ESP of the soil. The bio-resource *Sesbania rostrata* applied @ 25 per cent N equivalent recorded the lowest pH of 8.15, EC of 1.22 dS m^{-1} and ESP of 17.95 at the end of the N incubation experiment. Application of enriched FYM for 25 per cent P substitution recorded the lowest pH of 8.21, EC of 1.28 dS m^{-1} and ESP of 19.75 in the P incubation experiment. The application of composted coirpith for the partial substitution of N and *Casuarina* needles for the partial substitution of P also had significant influence in reducing the salinity and sodicity of the coastal salt affected soil. In general, all the bio-resources showed reduced pH, EC and ESP with the time of incubation.

The favorable reduction in the soil salinity and sodicity in both the incubation experiments could be attributed to the prolonged decomposition of added bio-resources. During the process of decomposition, a significant quantity of CO_2 is liberated and the decomposed product contains appreciable amounts of organic acids (Anand Swarup, 1992; Baziramakenga and Simmard, 1998) that consequently contributed to the reduction in the physico-chemical properties. The mineralization process during the decomposition of applied bio-resources might have also favoured the improvement in the physico-chemical properties. Anand Swarup (1988), Savithri *et al.* (1991), Rathore *et al.* (2004) Singaravel and Prasath (2004) and Sultan Singh *et al.* (2005) also made similar observations.

Table 1. Effect of bio-resource treatments on the physico-chemical properties and ESP of the soil in nitrogen incubation experiment

Treatments	pH			EC (dS m^{-1})			ESP		
	30 DAI	60 DAI	90 DAI	30 DAI	60 DAI	90 DAI	30 DAI	60 DAI	90 DAI
T_1	8.85	8.80	8.78	1.79	1.69	1.60	29.94	28.10	27.77
T_2	8.83	8.79	8.74	1.78	1.67	1.53	28.70	28.64	27.38
T_3	8.27	8.20	8.15	1.58	1.45	1.22	23.30	19.02	17.95
T_4	8.56	8.52	8.45	1.66	1.58	1.38	28.57	26.87	25.56
T_5	8.48	8.40	8.33	1.64	1.47	1.24	25.74	21.60	20.41
T_6	8.62	8.41	8.34	1.55	1.48	1.32	23.65	19.15	18.02
T_7	8.64	8.58	8.53	1.67	1.61	1.49	27.74	26.80	25.39
T_8	8.63	8.53	8.42	1.54	1.49	1.33	27.11	24.69	22.51
T_9	8.53	8.47	8.44	1.56	1.48	1.37	24.01	23.61	19.73
T_{10}	8.61	8.57	8.52	1.67	1.59	1.45	27.60	27.61	25.35
T_{11}	8.59	8.51	8.43	1.57	1.50	1.37	26.11	23.17	22.17
SEd	0.09	0.10	0.09	0.03	0.04	0.03	1.12	1.04	0.93
CD (p=0.05)	0.18	0.19	0.17	0.06	0.08	0.05	2.23	2.07	1.86

DAI-Days After Incubation, T_1-Absolute control, T_2-100% NPK, T_3- 75% N + 25% N through *Sesbania rostrata* + PK, T_4- 75% N + *Azospirillum* + PK, T_5- 75% N + 25% N through composted coirpith + PK, T_6- 50% N + 50% N through *Sesbania rostrata* + PK, T_7- 50% N + *Azospirillum* + PK, T_8- 50% N + 50% N through composted coirpith + PK, T_9- 25% N + 75% N through *Sesbania rostrata* + PK, T_{10}- 25% N + *Azospirillum* + PK, T_{11}- 25% N + 75% N through composted coirpith + PK

Table 2. Effect of bio-resource treatments on the physico-chemical properties and ESP of the soil in phosphorus incubation experiment

Treatments	pH			EC (dS m^{-1})			ESP		
	30 DAI	60 DAI	90 DAI	30 DAI	60 DAI	90 DAI	30 DAI	60 DAI	90 DAI
T_1	8.95	8.93	8.90	1.87	1.80	1.78	32.71	31.25	30.78
T_2	8.92	8.89	8.86	1.83	1.80	1.46	31.18	31.02	30.56
T_3	8.59	8.38	8.21	1.68	1.55	1.28	26.50	22.64	19.75
T_4	8.75	8.60	8.50	1.77	1.73	1.66	29.75	28.17	26.21
T_5	8.66	8.46	8.38	1.64	1.61	1.55	28.74	25.30	23.54
T_6	8.72	8.61	8.54	1.76	1.70	1.61	27.44	26.36	24.17
T_7	8.79	8.66	8.64	1.80	1.78	1.71	29.81	28.20	27.92
T_8	8.75	8.66	8.60	1.78	1.74	1.69	28.76	26.59	25.14
T_9	8.73	8.57	8.51	1.76	1.69	1.61	27.38	26.45	24.77

T_{10}	8.76	8.68	8.50	1.82	1.80	1.74	29.65	28.15	27.83
T_{11}	8.75	8.67	8.51	1.79	1.75	1.70	28.56	26.34	25.68
SEd	0.04	0.03	0.03	0.04	0.03	0.02	0.99	0.87	0.94
CD (p=0.05)	0.08	0.06	0.05	0.07	0.05	0.04	1.97	1.74	1.87

T_1.Absolute control, T_2. 100% NPK, T_3. 75% P + 25% P through enriched FYM + NK, T_4. 75% P + *Phosphobacterium* + NK, T_5. 75% P + 25% P through Casuarina needles + NK, T_6. 50% P + 50% P through enriched FYM + NK, T_7. 50% P + *Phosphobacterium* + NK, T_8. 50% P + 50% P through Casuarina needles + NK, T_9. 25% P + 75% P through enriched FYM + NK, T_{10}. 25% P + *Phosphobacterium* + NK, T_{11}. 25% P + 75% P through Casuarina needles + NK

Pot and Field Experiments

One pot experiment was conducted with selected bio-resource treatments from the incubation experiments which proved more efficient in improving the physico-chemical properties and nutrient availability.

For the pot experiment, 75 per cent N & P through chemical fertilizers and bio-resources, green manure & *Azospirillum* for 25 per cent N substitution & enriched FYM & *Phosphobacterium* for 25 per cent P substitution & their combinations were evaluated. In a Completely Randomised Design with three replications, the following treatments: T_1 - Control (100% NPK), T_2 - 75% N + 25% N through *Sesbania rostrata* + PK, T_3 - 75% N + *Azospirillum* + PK, T_4 - 75% P + 25% P through enriched Farm yard manure + NK, T_5 - 75% P + *Phosphobacterium* + NK, T_6 - 75% N + 25% N through *Sesbania rostrata* + 75% P + 25% P through enriched Farm yard manure + K, T_7 - 75% N + 25% N through *Sesbania rostrata* + 75% P + *Phosphobacterium* + K, T_8 - 75% N + *Azospirillum* + 75% P + 25% P through enriched Farm yard manure + K, T_9 - 75% N + *Azospirillum* + 75% P + *Phosphobacterium* + K were evaluated.

To verify the validity of the various bio-resources which proved very efficient in increasing the soil fertility and yield of rice in pot experiment under field conditions, experiments were carried out using rice using var.ADT 36 under field condition in the farmer's locale The field experiments were conducted in randomised block design (RBD) with four replications involving the following five treatments: T_1- Control (100% NPK), T_2- 75% NP + 25% N through *Sesbania rostrata* + 25% P through Super Phosphate enriched FYM + K, T_3- 75% NP + 25% N through *Sesbania rostrata* + *Phosphobacterium* + K, T_4- 75% NP + *Azospirillum* + 25% P through Super Phosphate enriched FYM + K, T_5- 75% NP + *Azospirillum* + *Phosphobacterium* + K. Fertilizer application was followed at the rate of 120: 38: 38 kg of N: P_2O_5: K_2O for rice. At critical stages of rice soil samples were collected and analysed for the physico-chemical characteristics. The results obtained are discussed hereunder with suitable literatures.

Soil reaction (pH)

The application of various bio-resources helped in reducing the pH of the coastal salt affected soil as evidenced in both the pot (Table 3) and field experiments (Table 4&5). In the pot experiment, the reduction in pH with bio-resources ranged from 8.09 to 8.63 at the harvest of the crop and ranged from 8.13 to 8.75 in the first year and from 7.92 to 8.66 in the second year of field experiment. Among the bio-resources treatments, the application of green manure and enriched FYM excelled all others in reducing the pH of the soil. The reduction was more pronounced in the second year of rice cultivation.

The decrease in soil pH may be attributed to the higher production of CO_2 and organic acids (Wen Qixiao and Tu Tianren, 1988; Anand Swarup, 1992) during the decomposition of the applied bio-resources. In the flooded soils, the production of CO_2 and its effect on redox reactions significantly altered in soil pH (Ponnamperuma, 1977). Yaduvanshi (2001), Prasad *et al.* (2002) and Bharat Singh and Pandey (2004) also reported the reduction in soil pH due to the addition of various bio-resources.

Table 3. Effect of bio-resource treatments on the physico-chemical properties and ESP of the soil in the pot experiment

	pH			EC (dS m^{-1})			ESP		
Treatments	**AT**	**PI**	**Har**	**AT**	**PI**	**Har**	**AT**	**PI**	**Har**
T_1	8.84	8.79	8.79	1.66	1.63	1.61	28.83	28.20	27.94
T_2	8.60	8.41	8.28	1.44	1.40	1.32	25.07	22.67	21.30
T_3	8.74	8.67	8.63	1.60	1.57	1.55	27.97	27.60	27.20
T_4	8.64	8.53	8.35	1.51	1.47	1.40	25.60	24.50	22.53
T_5	8.73	8.67	8.62	1.60	1.57	1.52	28.07	27.70	27.30
T_6	8.47	8.23	8.09	1.42	1.24	1.12	23.43	21.27	20.13
T_7	8.55	8.35	8.23	1.52	1.32	1.29	24.97	23.07	20.87
T_8	8.58	8.41	8.36	1.46	1.41	1.37	24.83	23.17	22.10
T_9	8.65	8.62	8.57	1.59	1.55	1.51	27.37	26.93	26.53
SEd	0.04	0.04	0.03	0.04	0.04	0.03	0.38	0.46	0.34
CD (p=0.05%)	0.08	0.07	0.06	0.07	0.07	0.06	0.78	0.92	0.63

AT – Active Tillering stage; PI - Panicle Initiation stage; Har - Harvest stage, T_1 - Control (100% NPK), T_2 - 75% N + 25% N through *Sesbania rostrata* +PK, T_3 - 75% N + *Azospirillum* + PK, T_4 - 75% P + 25% P through enriched Farm yard manure + NK, T_5 - 75% P + *Phosphobacterium* + NK, T_6 - 75% N + 25% N through *Sesbania rostrata* + 75% P + 25% P through enriched Farm yard manure + K, T_7 - 75% N + 25% N through *Sesbania rostrata* + 75% P + *Phosphobacterium* + K, T_8 - 75% N + *Azospirillum* + 75% P + 25% P through enriched Farm yard manure + K, T_9 - 75% N + *Azospirillum* + 75% P + *Phosphobacterium* + K

Table 4. Effect of bio-resource treatments on the physico- chemical properties, ESP and CEC of the soil in the field experiment-I

	pH			**EC (dS m^{-1})**			**ESP**			**CEC (C mol (P^{+}) kg^{-1}**		
Treatments	**AT**	**PI**	**Har**	**AT**	**PI**	**Har**	**AT**	**PI**	**Har**	**AT**	**PI**	**Har**
T$_1$	8.86	8.81	8.78	1.54	1.50	1.48	29.4	29.2	28.5	17.4	17.4	17.4
T$_2$	8.32	8.23	8.13	1.32	1.25	1.21	23.3	20.4	18.6	17.8	17.9	18.2
T$_3$	8.41	8.29	8.17	1.37	1.29	1.25	26.4	24.2	21.5	17.0	17.8	18.0
T$_4$	8.63	8.56	8.42	1.41	1.35	1.31	27.4	26.4	23.2	17.6	17.7	17.8
T$_5$	8.80	8.78	8.75	1.50	1.46	1.45	29.2	28.5	27.5	17.4	17.4	17.4
SEd	0.14	0.16	0.11	0.01	0.01	0.02	0.32	1.26	1.04	0.22	0.18	0.19
CD (p=0 .05%)	0.27	0.32	0.21	0.02	0.02	0.04	0.64	2.53	2.08	0.43	0.36	0.37

T$_1$- Control (100% NPK), T$_2$- 75% NP + 25% N through *Sesbania rostrata* + 25% P through Super Phosphate enriched FYM + K, T$_3$- 75% NP + 25% N through *Sesbania rostrata* + *Phosphobacterium* + K, T$_4$- 75% NP + *Azospirillum* + 25% P through Super Phosphate enriched FYM + K, T$_5$- 75% NP + *Azospirillum* + *Phosphobacterium* + K

Table 5. Effect of bio-resource treatments on the physico chemical properties, ESP and CEC of the soil in the field experiment-II

	pH			**EC (dS m^{-1})**			**ESP**			**CEC (C mol (P^{+}) kg^{-1}**		
Treatments	**AT**	**PI**	**Har**	**AT**	**PI**	**Har**	**AT**	**PI**	**Har**	**AT**	**PI**	**Har**
T$_1$	8.79	8.70	8.68	1.45	1.41	1.38	28.5	28.0	27.8	17.4	17.5	17.5
T$_2$	8.07	8.01	7.92	1.16	1.09	1.02	17.2	15.5	14.8	18.2	18.5	18.7
T$_3$	8.11	8.06	8.00	1.21	1.16	1.12	20.4	18.8	15.1	18.1	18.2	18.4
T$_4$	8.36	8.31	8.25	1.26	1.21	1.15	22.5	20.7	19.0	17.9	18.0	18.1
T$_5$	8.73	8.70	8.66	1.40	1.37	1.34	27.1	26.5	24.8	17.4	17.5	17.6
SEd	0.21	0.23	0.21	0.03	0.03	0.04	0.53	0.64	0.59	0.09	0.12	0.08
CD (p=0 .05%)	0.42	0.35	0.31	0.05	0.06	0.07	1.06	1.27	1.18	0.18	0.23	0.15

Electrical conductivity (EC)

A favorable reduction in electrical conductivity was observed in the pot (Table 3) and field experiments (Table 4&5) due to the application of various bio-resources. In the pot experiment, the EC was reduced to the tune of 1.12 to 1.55 dS m^{-1} in the bio-resources treated plots. In the field experiments, the reduction was in the range of 1.05 to 1.34 dS m^{-1} at the end of second year of rice cultivation. The highest reduction was found with the application of treatment, 75% NP + green manure + enriched FYM. In the earlier reports, Malik *et al.*

(2004) and Kaushik *et al.* (2005) observed the lowering of EC and the alleviation of salt effects due to the application of phytomass and FYM amendments in salt affected soils.

Application of bio-resources improves the leaching of salts, sequesters salts and thus exerts ameliorative action (Singh and Roy, 1975). The another probable reason for lowering of EC could be the grater leaching rate of soluble salts from the amended bio- resources to the soil than the supplied rate and also leaching of salts by weathering (Wang and Yang, 2003).

Exchangeable sodium percentage (ESP)

The exchangeable sodium percentage of the coastal saline sodic soil significantly decreased due to the application of various bio-resources (3,4&5). Among the treatments, the soil incorporation of green manure and enriched FYM as the sources for 25 per cent N and P respectively excelled over other treatments in reducing the ESP of the soil. This treatment recorded 20.1 per cent of exchangeable sodium at the harvest stage of pot experiment, 18.6 per cent at the harvest of first field experiment and 14.8 per cent at the harvest of second field experiment respectively.

The submergence of sodic soil leads to the accumulation of CO_2, which solubilizes the native $CaCO_3$ and Ca thus released in turn replaces the exchangeable Na from the exchange complex (Anand Swarup, 1992). In the present study, the application of green manure and enriched FYM was more effective in reducing the ESP than all other treatments. This was mainly because green manure + enriched FYM generated more CO_2 and also they contain appreciable amount of Ca that could be available in greater amounts for exchange with Na ions. Further, the enhanced microbial activities accompanied with liberation of CO_2 and other bio chemical reactions might have acted in the process of reducing the ESP of soil.

Kausik *et al.* (2005) suggested altogether a different concept that explains the reduction of ESP in the soil. According to them, the incorporation of bio-resources resulted in the binding or chelation of the cation (Na) to the organo-colloids formed by the decomposition of biomass, thereby decreasing the ESP.

Bharambe *et al.* (2001), Kaushik *et al.* (2005) and Sultan Singh *et al.* (2005) were also reported the reduction in soil ESP due to the application of various bio-resources.

Cation exchange capacity (CEC)

The CEC of the post harvest soils in both the cropping year showed a marginal increase with the application of bio-resource treatments (Table 4&5). Among the bio-resource treatments, the soil incorporation of green manure + enriched FYM registered a CEC of 18.0 C mol (p^+) kg^{-1} in the first year and 18.7 C mol (p^+) kg^{-1} in the second year.

The beneficial effect of applied bio-resources in increasing the CEC of soil could be due to the continued decomposition of biomass and release of humic compounds, which are known for increasing cation exchange capacity (Mays *et al.*, 1973). Similar results of increased CEC in the soil due to the application of various kind of bio-resources were earlier reported by Neelam Sharad and Verma (2001), Bullock III *et al.* (2002) and Prakash *et al.* (2002).

Physical properties

In the present investigation, an improvement in the soil physical properties due to the application of various bio-resources was well evidenced (Table 6). The effect was more pronounced in the second year of rice cultivation.

Table 6. Effect of bio-resource treatments on the physical properties of the soil in the field experiment- I

	Field experiment 1			Field experiment 2		
Treatments	Bulk Density ($Mg\ m^{-3}$)	Particle Density ($Mg\ m^{-3}$)	Per cent Porosity	Bulk Density ($Mg\ m^{-3}$)	Particle Density ($Mg\ m^{-3}$)	Per cent Porosity
T_1	1.40	1.89	51.08	1.39	1.88	52.62
T_2	1.30	1.86	56.25	1.27	1.85	58.60
T_3	1.32	1.88	56.00	1.29	1.86	58.18
T_4	1.34	1.88	53.78	1.33	1.88	54.50
T_5	1.36	1.89	52.15	1.34	1.88	53.72
SEd	0.02	0.04	0.75	0.02	0.05	1.32
CD (p=0.05)	0.03	NS	1.49	0.03	NS	2.63

At the end of two years of rice cultivation, the soil bulk density reduced to a tune of 1.27 to 1.34 Mg m^{-3} with various bio-resources application. The particle density ranged from 1.85 to 1.88 Mg m^{-3} with out any significant difference due to bio- resources application. The percent porosity of the soil was also significantly increased with the bio-resources application. At the harvest of the second crop, the porosity increased to a range of 53.72 to 58.60 per cent due to the bio-resources application as against the control, which registered 52.62 per cent.

Generally in rice soils the decrease in bulk density with the advancement of rice crop growth are due to resettling of soil particles and over burden pressure of ponded water (De Dutta and Kerim, 1974). The bio-resources application had an added advantage in reducing the bulk density of the soil. The soil organic carbon plays an important role in cementing the soil aggregates as well as promoting the soil structure (Wang and Yang, 2003). It is well-established fact that the bulk density depends on soil structure (Celik *et al.*, 2004). The decreased bulk density due to the application of various bio-resources by the way of supplying higher organic matter was reported by many authors (Tiarks, 1974; Paul and Clark, 1996; Nyakatawa *et al.*, 2001).

The bio-resources had positive effect on the percent pore space of the soil. Aggelider and Loadra (2000) demonstrated that the addition of organic matter considerably improved the soil physical properties by increasing total porosity and changing distribution of pore sizes in loamy and clay-textured soils. Marinari *et al.* (2000) also found that the total soil porosity increased with organic fertilizers and compost, depending on the amount of materials added.

Celik *et al.* (2004), Mandal *et al.* (2003), Manna *et al.* (2005) and Selvi *et al.* (2005) were also reported the improved soil physical properties due the addition of various bio- resources in the soil.

3

Bioresources and Soil Nitrogen

It is generally observed that N, P and Zn content in soil decreases with increasing soil salinity. Nutrient imbalances created by continuous use of chemical fertilizers, particularly nitrogenous fertilizer alone, combined with sub optimal rates of other nutrients or total omission of other nutrients are the primary causes for unsustainable yields of crops in saline soil. A sustainable approach for nutrient management in salt affected soils should therefore include balanced use of organic, inorganic and biological sources of nutrients (Bandyopadhyay and Rao, 2001). Organic amendments are known to facilitate the reclamation of sodic soils and better nutrition of crops (Anand swarup, 1992).

Organic carbon

In most of the environmental conditions, soil organic matter decreases faster than its replenishment under continuous farming without manures (Dalal and Mayer, 1986; Golchin *et al.*, 1993; Marinari *et al.*, 2000). It is well established that addition of organic materials in soil improves the organic carbon content of the soil, which is important as it governs the physical (Celik *et al.*, 2004), chemical and biological (Marinari *et al.*, 2000) properties of the soil.

Experimental evidences showed that long-term application of fertilizers and organic manures under continuous cropping improves soil organic carbon content (Acharya *et al.*, 1988).

Bhandari *et al.* (1992) found that organic carbon content of the soil increased significantly with the application of 100 per cent RDF and the conjugative use of organic manures viz. FYM, and green manures. Mayalagu *et al.* (1995) reported that application of coirpith @ 60 t ha^{-1} increased the organic carbon content, besides improving the NPK status of the soil.

Bellakki *et al.* (1998) observed the highest organic carbon content (5.6 g kg^{-1}) with the incorporation of *glyricidia* along with 50 per cent NPK, which was significantly higher than various other fertilizer levels. Jane *et al.* (1999) reported that the application of cattle manure @ 10 t ha^{-1}along with 100 per cent recommended dose of fertilizers (RDF) and maize stover in the soil resulted in the least loss of organic carbon under a long-term field experiment.

Babhulkar *et al.* (2000) noticed that the organic carbon content of soil significantly increased due to application of FYM along with inorganic fertilizers. Similar results were also reported earlier by Babu and Reddy (2000). Yadav *et al.* (2000) found that the application of 50 percent of recommended dose of nitrogen (RDN) + 50 per cent N through FYM or through *Sesbania sesban* to rice improved the organic carbon content of the soil over other inorganic inputs.

Mandal *et al.* (2003) reported a significant increase in the organic carbon content of the soil due to the addition of green manures. *Sesbania rostrata* was found superior than *Sesbania aculeata* in increasing the organic carbon content of the soil. Manna *et al.* (2005) revealed that the soil organic carbon considerably increased with the application of 100 per cent RDF and FYM @ 10 Mg ha^{-1}.

Earlier experiments results indicated that total N, available P and K content of soils were improved by the combined use of organic and inorganic sources of nutrients as compared to use of inorganic form alone (Nambiar and Abrol, 1992; Badanur and Tolanur, 1992; Babu and Reddy, 2000; Sharma *et al.*, 2001; Saleque *et al.*, 2004)

Singh *et al.* (2006) reported that the content of organic carbon and available phosphorus in soil improved in all integrated nutrient-management practices namely 25% RDN through pressmud + 75% RDF through inorganic fertilizers, and 50% RDN through pressmud + 50% RDN through inorganic fertilizers compared with its initial soil status.

Baishya *et al.* (2015) reported that the plot receiving 2.5 t poultry manure ha^{-1} + 125% CDF registered highest available organic carbon, N, P and K status in the soil. The application of organic manures along with inorganic fertilizers increased soil organic carbon (7.8-9.4 g kg^{-1}) nitrogen (166.6-203.0 kg ha^{-1}), phosphorus (23.3-32.6 kg ha^{-1}) and potassium (61.0- 92.3 kg ha^{-1}).

Han *et al.* (2018) reported that straw incorporation significantly sequestered soil organic carbon @ 0.35 Mg C ha^{-1} $year^{-1}$, the straw incorporation at 3 Mg C ha^{-1} yr^{-1} with the mineral N rate of 200-400 kg N ha^{-1} $year^{-1}$ was found to be the best combination for farmer use.

Velmurugan and Swarnam (2021) investigated the effects of enriched manures, inorganic fertilizers and their integrated use on maize-rice cropping system. Due to integrated use of enriched manures (*Azospirillum*, P solubilizer and growth promoter) and chemical fertilizers *i.e.* 50% NPK + enriched compost (3.5 t h^{-1}) a significant increase in labile carbon (869 mg kg^{-1}), MBC (290 mg kg^{-1}) and mineralizable carbon (12.20 µg CO_2 g^{-1} h^{-1}) in rice soils was recorded at post-harvest stage.

Nitrogen

Alkali soils are generally low in humus and thus nitrogen content. The available N content decreases with increasing soil pH. Besides the low nitrogen content, alkalinity adversely affects N-transformations leading to nitrite accumulation, volatilization and denitrification of the added nitrogen (Tiwari and Sharma, 1989). Green manuring with different plant species is very useful in alkali soils. Application of 40 kg N ha^{-1} along with green manuring gives rice equivalent to 120 kg N ha^{-1} without green manuring thus, there could be saving of 80 kg N ha^{-1} (Tiwari *et al.*, 1980).

Clarson *et al.* (1984) revealed that application of composted coirpith increased the soil properties like CEC and available N, P and K in the soil. Selvi and Augustine Selvaseelan (1991) reported that application of composted coirpith in the soil increased almost all the nutrients considerably as compared to the application of chemical fertilizers alone. Continuous addition of FYM and sunhemp either alone or in combination with inorganic fertilizers recorded significantly higher available nitrogen and phosphorus than control (Bellakki and Badanur, 1997).

Long term experiments on integrated nutrient supply system in rice- rice sequence indicated that the available N content of soil increased significantly under integrated treatments receiving both inorganic and organic sources of fertilizers. Available P also showed an improvement with the addition of 75 and 100 per cent NPK and their combined use with different organic sources (Basumatary and Talukdar, 1998)

Sheeba and Chellamuthu (1999) reported that application of FYM in conjugation with 100 per cent RDF registered highest total nitrogen, available phosphorus and potassium contents in an inceptisols. Paramanik and Mahapatra (1997) found that application of 50 per cent N through prilled urea and remaining substituted through *Sesbania aculeata* resulted in the highest concentration of NH_4- N in the soil over other N sources. Organic carbon and available N content of soil in jute- rice cropping sequence increased appreciably by the application of organic materials through water hyacinth, FYM and paddy straw when compared to chemical fertilizers alone (Singh and Ghosh, 1999).

Sharma *et al.* (2000) demonstrated through a four-year experiment that incorporation of *Sesbania aculeata* increased the available nitrogen and phosphorus in a rice- wheat cropping system. The earlier report of Vinodkumar *et al.* (1998) also indicates increased N status of the soil due to the incorporation of green manures. Gupta *et al.* (2003) stated that higher concentration of mineral-N was found in the soil with the application of biogas slurry or FYM along with inorganic fertilizers. Sonune *et al.* (2003) reported that 36 per cent increase in available nitrogen by the addition of manures and plant biomass, which on mineralization enhanced the nitrogen pool in the soil.

Application of 100% recommended dose of N and green manuring with *Sesbania aculeata* increased the available nitrogen and potassium. The nutrient balance remained positive for these two nutrients after two successive rice crops (Selvi *et al.* 2003a). Masciandaro *et al.* (2004) revealed that application of dairy manure compost @ 35 and 70 t ha^{-1} in the soil increased the NH_4^+- N and NO_{3-}^- - N in the soil.

Rama Lakshmi *et al.* (2012) through field experiment during kharif and rabi 2009 and 2010 reported that the nitrogen use efficiency increased with the application of 75 % chemical fertilizers + weed vermicompost @ 2.5 t ha^{-1} was better than other combinations closely followed by 50% RDF + 50% prathista organics. Kyi Moe *et al.* (2019) concluded that the organic fertilizer (total N ≥ 4%) with the EMN method enhances higher N availability in each year. Continuous application of organic fertilizer (total N < 4%) over two years effectively increased N availability in the second year. The 50% organic fertilizer (total N ≥ 4%) and 50% CF led to increased NPK availability and rice yields over the 100% CF treatment, reducing CF usage and leading for sustainable agriculture.

Gogoi *et al.* (2018) revealed that the highest soil available 293.82 N kg ha^{-1}, 66.91 P kg ha^{-1} and 146.34 K kg ha^{-1} content were recorded with enriched compost @ 5t ha^{-1} applied fields.

An incubation experiment on N was conducted with treatments constituting eleven viz., absolute control, 100 per cent NPK, various bio- resources *viz. Sesbania rostrata*, composted coirpith and *Azospirillum* as sources of N substitution for 25, 50 and 75 percent of the N dose for rice crop (120 kg ha^{-1}) as listed in Table 1. For the phosphorus incubation experiment, eleven treatments involving absolute control and 100 per cent NPK, the bio-resources *viz.* super phosphate enriched farmyard manure, *Casuarina* needles and *Phosphobacterium* were used for supplementing 25, 50 and 75 percent of the recommended P for rice crop (38 kg ha^{-1}). The treatment details are given in Table 2. In both the incubation experiments, periodic soil samples at 30, 60 and 90 days after incubation (DAI) were taken and analysed for the organic carbon and Alkaline $KMnO_4$-N. The results obtained were discussed in line with suitable literatures.

Organic carbon

The organic carbon content of the experimental soil was 0.24 per cent and rated low. The organic carbon content of the incubated soil was greatly influenced by the added bio-resources as evidenced in the two incubation experiments (Table 1&2). The increase in the organic carbon content of the soil was in the range of 0.29 to 0.60 per cent in the first incubation experiment and 0.30 to 0.56 per cent in the second incubation experiment due to the application of various bio-resources. Among the various bio-resources, the green manure applied as the substitution for 25 per cent N and the enriched FYM substituted for 25 per cent P in the first and second incubation experiments respectively excelled all other bio-resources by recording the organic carbon content of 0.60 and 0.56 per cent respectively.

The reduced pH and EC in both the incubation experiments might have helped for better proliferation of soil microbes and enhanced the decomposition of added bio-resources and in turn the organic carbon status. The bio-resources used in these incubation studies carrying significant carbon, had low C: N ratio (15.07 in *Sesbania rostrata*, 23.10 in composted coirpith, 21.51 in enriched FYM and 23.45 in *Casuarina* needles) which might have hastened the microbial decomposition and subsequently build up of the organic carbon status in the soil (Sud *et al.* 1990). Bellakki *et al.* (1998), Babhulkar *et al.* (2000), Mandal *et al.* (2003) and Manna *et al.* (2005) already reported the increased organic carbon content of the soil due to the application of bio-resources.

Alkaline $KMnO_4$- N

Salt affected coastal soils are in general show low nutrient status especially nitrogen and phosphorus, while in most cases K is well supplied. As advocated by many workers (Nambiar and Abrol, 1992; Bharadwaj and Omanwar, 1994; Sharma *et al.*, 2001; Saleque *et al.*, 2004) a judicious integration of bio-resources with inorganic fertilizers will improve and sustain the nutrient availability of these soils.

In both the incubation experiments, the availability of N increased with the application of various bio-resources (Table 1&2). In the N incubation experiment, the application of 75% N through chemical fertilizers + 25% N substituted through green manure (151.82 mg kg^{-1} at 90 DAI) and 75 % N + *Azospirillum* inoculation (146.22 mg kg^{-1} at 90 DAI) registered the significantly highest N availability (Fig.1). With registering 167.50 and 155.28 mg kg^{-1} at 90 DAI, the application of 75% P and 25% P supplied through enriched FYM and 25% P through *Casuarina* needles respectively excelled in increasing the N availability in the phosphorus incubation experiment. The bio-resources composted coirpith and

Phosphobacterium inoculation also had positive influence on increasing the availability of N in the soil.

The increased N availability with applied bio-resources might be due to the increased decomposition of applied bio-resources under favorable soil environment and due to reduced volatilization, leaching and denitrifying losses (Clarson *et al.*, 1984). Duhan *et al.* (2005) suggested that the green-manure treated soils act as harbor of numerous microorganisms that are associated with fast decomposition and mineralization of organic materials in the soil and thus the availability of N greatly increased. The inoculation of *Azospirillum* in the soil lead to the increased growth and multiplication resulting in enhanced N_2 fixation. Rivarola *et al.* (1998) reported the salt tolerance behaviour of *Azospirillum* coupled with the submergence of incubated soil made the soil environment congenial for the growth of *Azospirillum*. Gupta and Narwal (1988) and Imagava (1989) reported the similar favorable results of increased N concentration in the incubated soil.

Table 1. Effect of bio-resource treatments on the organic carbon (%) and available nitrogen content (mg kg^{-1}) of soil in nitrogen incubation experiment

	Org. Carbon			**Alkaline $KMnO_4$ –N**		
Treatments	**30 DAI**	**60 DAI**	**90 DAI**	**30 DAI**	**60 DAI**	**90 DAI**
T_1	0.26	0.26	0.27	112.52	113.25	114.43
T_2	0.38	0.28	0.28	125.36	129.56	131.75
T_3	0.49	0.55	0.60	143.28	148.21	151.82
T_4	0.25	0.28	0.29	139.50	143.72	146.22
T_5	0.46	0.49	0.51	136.15	140.16	142.55
T_6	0.45	0.48	0.52	135.75	138.65	140.43
T_7	0.26	0.27	0.30	133.11	135.20	136.69
T_8	0.40	0.43	0.47	129.92	132.76	133.47
T_9	0.47	0.52	0.54	127.55	129.92	130.77
T_{10}	0.26	0.28	0.29	126.75	127.36	128.96
T_{11}	0.39	0.43	0.47	126.54	127.00	128.57
SEd	0.02	0.02	0.03	4.33	3.22	4.14
CD (p=0.05)	0.04	0.04	0.05	8.65	6.44	8.27

T_1-Absolute control, T_2-100% NPK, T_3- 75% N + 25% N through *Sesbania rostrata* + PK, T_4- 75% N + *Azospirillum* + PK, T_5- 75% N + 25% N through composted coirpith + PK, T_6- 50% N + 50% N through *Sesbania rostrata* + PK, T_7- 50% N + *Azospirillum* + PK, T_8- 50% N + 50% N through composted coirpith + PK, T_9- 25% N + 75% N through *Sesbania rostrata* + PK, T_{10}- 25% N + *Azospirillum* + PK, T11- 25% N + 75% N through composted coirpith + PK

Table 2. Effect of Bio-resource treatments on the organic carbon (%) and available nitrogen content (mg kg^{-1}) of soil in Phosphorus incubation experiment

	Org. Carbon			Alkaline $KMnO_4$ –N		
Treatments	**30 DAI**	**60 DAI**	**90 DAI**	**30 DAI**	**60 DAI**	**90 DAI**
T_1	0.26	0.27	0.28	118.00	119.57	119.65
T_2	0.27	0.28	0.28	125.90	127.63	129.43
T_3	0.38	0.47	0.56	149.18	158.74	167.50
T_4	0.29	0.29	0.31	129.63	137.72	140.72
T_5	0.33	0.41	0.49	141.25	150.25	155.28
T_6	0.36	0.43	0.47	137.10	144.74	147.56
T_7	0.29	0.30	0.30	131.30	138.27	140.50
T_8	0.29	0.37	0.40	129.63	137.25	140.17
T_9	0.35	0.42	0.48	134.57	143.65	145.28
T_{10}	0.28	0.29	0.30	126.13	135.27	138.25
T_{11}	0.30	0.36	0.38	128.73	138.45	141.36
SEd	0.02	0.02	0.03	3.85	3.20	4.64
CD (p=0.05)	0.03	0.04	0.05	7.70	6.40	9.27

T_1.Absolute control, T_2. 100% NPK, T_3. 75% P + 25% P through enriched FYM + NK, T_4. 75% P + *Phosphobacterium* + NK, T_5. 75% P + 25% P through Casuarina needles + NK, T_6. 50% P + 50% P through enriched FYM + NK, T_7. 50% P + *Phosphobacterium* + NK, T_8. 50% P + 50% P through Casuarina needles + NK, T_9. 25% P + 75% P through enriched FYM + NK, T_{10}. 25% P + *Phosphobacterium* + NK, T_{11}. 25% P + 75% P through Casuarina needles + NK

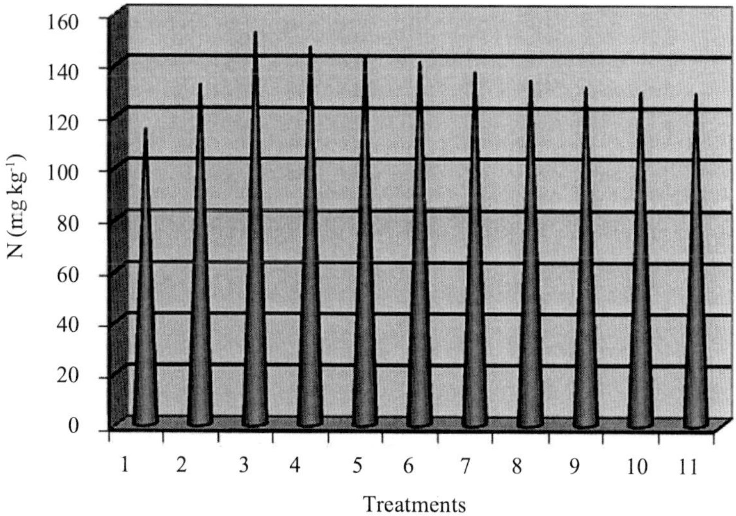

Figure 1. Effect of bio- resource treatments on the alkaline $KMnO_4$- N of the soil in N incubation experiment

Pot and field Experiment

Organic carbon

The organic carbon content of the soil, which is essential factor for sustaining the soil productivity, was significantly increased with the application of various bio-resources (Table 3). With the application of various bio-resources, the content of organic carbon in the soil increased to the tune of 0.26 to 0.53 percent in pot experiment and 0.27 to 0.49 per cent at the end of second field experiment as against 0.25 per cent in control. The application of 100 per cent NPK through fertilizers slightly increased the organic carbon content of the soil.

Several workers (Mishra and Sharma, 1997; Mandal *et al.*, 2003; Masciandaro *et al.*, 2004; Manna *et al.*, 2005) reported an increased organic carbon content of the soil due to the application of different kind of bio-resources.

Under most environmental conditions, soil organic carbon decreases with continuous farming without manure application (Oadas, 1984). Bio-resources are known to increase plant biomass yields due to their positive effects on physico- chemical and biological soil conditions. These effects include the better proliferation of rooting (Rathore *et al.*, 1998) and stimulation of biotic activity (Masciandaro *et al.*, 2004). The increased rooting and biomass production in turn add up organic matter to the soil (Bhandari *et al.*, 1992).

Furthermore, the added bio-resources contain appreciable amount of carbon. The increased biological activity will hasten the decomposition process of the added bio-resources and the resettling of the component organic C into the topsoil (Ballestero and Douglas, 1996; Eghball *et al.*, 1997) increased the organic carbon content of the soil. A significant positive correlation of microbes with organic carbon content of the soil (r = 0.98* and 0.96** for bacteria, r = 0.96** and 0.95** for fungi and r = 92** and 0.93** for actinomycetes in the first and second year of field experiment respectively) supports the present finding of the study. The fertilizers application also helped in building up of organic carbon content in the soil through its indirect effect. The increased plant growth and root biomass due to the fertilizers application added C into the soil (Mandal *et al.*, 2003).

Alkaline $KMnO_4$- N

Nitrogen, the kingpin of rice production increased in the soil with the application of various bio-resources (Table 3&4) . The results of pot and field experiments indicated that the application of 75 % NP + 25 % N substituted through *Sesbania rostrata* + 25 % P substituted through enriched FYM rated best in recording the highest N availability in the soil (Fig.). It registered 112.33 mg kg^{-1} in the pot experiment and 237.18 kg ha^{-1} at the end of second field experiment. The application of *Azospirillum* along with enriched FYM (227.25 kg ha^{-1} at the end of second crop) and *Phosphobacterium* (226.00 kg ha^{-1} at the end of second field experiment) also significantly increased the N availability in the soil compared to control.

In all the experiments, the application of 75 % N + 25 % N substituted through green manure recorded the higher N availability. This was presumably due to the direct influence of decomposition and mineralization of N from the green manure and indirectly due to the biomass N turnover known as priming effect (Jenkinson *et al.*, 1985; Azam, 1990). Higher N content of the green manure and the rapid minaralization of N from this green manure (Chen Lizhu, 1988; Thangaraju and Kanniyan, 1990) also contributed towards the higher N availability in the soil.

The increased retention of ammonialcal nitrogen on humus complex because of the additive effect of green manure and FYM and the release of extracellular nitrogenous substances (Singh Parihar *et al.*, 2003) also the reason for the increased N availability in the soil. In addition, the higher lignin content of the fresh green manure enhanced the incorporation of N into stabilized soil organic constituents and thus the N availability showed higher with the bio-resources application.

The increased N availability with *Azospirillum* inoculation might be due to the fixation of atmospheric nitrogen in cropped lands (Yoav Bashan, 2004). Earlier literatures also indicate the *Azospirillum* survival in saline environment and their possible role in N_2 fixation. *Azospirillum* has remarkable tolerance for saline condition induced osmatic stress (Rivarola *et al.*, 1998). The common cellular mechanism of osmatic stress adaptation is intercellular accumulation of organic solutes (osmolytes) such as glysine betaine, glutamate, proline and trehalose to adapt to fluctuarions in soil salinity / osmolarity.

Sharma *et al.* (2000), Mandal *et al.* (2003), Selvi *et al.* (2003b) and Masciandaro *et al.* (2004) were also reported the increased N concentration in the soil due to the application of various bio-resources.

Table 3. Effect of bio-resource treatments on the organic carbon (%) and available nitrogen (mg kg^{-1}) content of the soil in the pot experiment

Treatments	Org. C			N		
	AT	PI	Har	AT	PI	Har
T_1	0.24	0.24	0.25	103.66	98.46	95.25
T_2	0.35	0.40	0.42	114.28	109.33	105.76
T_3	0.25	0.26	0.26	109.86	105.27	102.00
T_4	0.33	0.37	0.37	107.56	102.76	99.86
T_5	0.25	0.26	0.27	109.47	100.21	97.54
T_6	0.40	0.46	0.53	123.56	117.66	112.33
T_7	0.33	0.38	0.44	119.71	114.82	108.56
T_8	0.32	0.38	0.39	110.56	106.76	103.76
T_9	0.26	0.26	0.29	111.42	107.23	104.82
SEd	0.02	0.02	0.02	1.81	2.01	1.59
CD (p=0.05%)	0.03	0.04	0.03	3.62	4.02	3.17

T_1 - Control (100% NPK), T_2 - 75% N + 25% N through *Sesbania rostrata* +PK, T_3 - 75% N + *Azospirillum* + PK, T_4 - 75% P + 25% P through enriched Farm yard manure + NK, T_5 - 75% P + *Phosphobacterium* + NK, T_6 - 75% N + 25% N through *Sesbania rostrata* + 75% P + 25% P through enriched Farm yard manure + K, T_7 - 75% N + 25% N through *Sesbania rostrata* + 75% P + *Phosphobacterium* + K, T_8 - 75% N + *Azospirillum* + 75% P + 25% P through enriched Farm yard manure + K, T_9 - 75% N + *Azospirillum* + 75% P + *Phosphobacterium* + K

Table 4. Effect of bio-resource treatments on the organic carbon (%) and available nitrogen (kg ha^{-1}) content of the soil in the field experiment-I

Treatments	Field Experiment 1						Field Experiment 2					
	Org. C			N			Org. C			N		
	AT	PI	Har	AT	PI	Har	AT	PI	Har	AT	PI	Har
T_1	0.24	0.25	0.25	225	212	208	0.25	0.25	0.26	219	205	201
T_2	0.32	0.41	0.49	249	239	232	0.48	0.52	0.57	255	246	237
T_3	0.29	0.37	0.42	241	232	227	0.41	0.47	0.50	244	237	232
T_4	0.27	0.29	0.32	232	226	221	0.35	0.38	0.42	237	229	227
T_5	0.25	0.26	0.27	227	212	215	0.28	0.29	0.29	234	228	226
SEd	0.01	0.01	0.02	6.0	3.7	4.3	0.02	0.02	0.02	4.2	3.9	4.0
CD (p=0.05%)	0.02	0.02	0.04	12.2	7.58	8.7	0.03	0.03	0.04	8.43	7.8	8.1

T_1- Control (100% NPK), T_2- 75% NP + 25% N through *Sesbania rostrata* + 25% P through Super Phosphate enriched FYM + K, T_3- 75% NP + 25% N through *Sesbania rostrata* + *Phosphobacterium* + K, T_4- 75% NP + *Azospirillum* + 25% P through Super Phosphate enriched FYM + K, T_5- 75% NP + *Azospirillum* + *Phosphobacterium* + K

4

Bioresources and Soil Phosphorus

Experiments at Central Soil Salinity Research Institute (CSSRI), Karnal, India showed that alkali soils are not always high in available P and rice and wheat significantly responded to addition of P (Pathak *et al.*, 1975).

The availability of phosphorus in soil either native or added is limited by fixation and precipitation mechanisms. The organic acids released during decomposition of organic matter influences the soil pH, form stable complexes with cations responsible for P fixation and thus increases the availability (Tiwari *et al.*, 1980).

The use of bulky organic manures (Maurya and Ghosh, 1972; Chaudhary *et al.*, 1981) and green manures (Prabhakar *et al.*, 1972; Sharma *et al.*, 2000) has been found very beneficial for P management of saline and alkali soils as these improve the soil physical condition and the availability of soil phosphates.

The addition of organic matter along with phosphatic fertilizers has been found to increase the availability of applied as well as native phosphorus (Subramanian and Gopalaswamy, 1991). Santhi *et al.* (1991) showed that combined application of composted coirpith @ 10 t ha^{-1} with 75 percent N increased available P and K by 30 and 44 per cent respectively. Patel *et al.* (1992) found that the addition of organic matter increased the availability of Olsen's P in the soil. Bhandari *et al.* (1992) reported a significant increase in available P due to the application of *Sesbania aculeata* as 50 per cent substitution of nitrogen.

Anand Swarup (1992) stated that there was an improvement in available P in the sodic soils treated with *Sesbania aculeata* and that could be attributed to the great mobilization of native soil phosphorus compounds and contribution through added biomass. Mayalagu *et al.* (1995) reported that application of coirpith @ 60 t ha^{-1} increased the availability of N, P and K content of the soil.

Santhi *et al.* (1998) observed a build up in the available phosphorus content of the soil over the initial value due to application of NPK along with FYM. Sheeba and Chellamuthu (1999) found that application of FYM along with RDF increased the available P status of an Inceptisol. Sharma *et al.* (2000) also reported the beneficial influence of FYM addition in relation to the availability of available P in the soil. Ravichandran (2003) revealed that addition of FYM or

Phosphobacteria or both resulted in significantly higher availability of available P in the soil.

Selvi *et al.* (2003a) stated that application of green leaf manure (*Pongamia glabra*) @ 6.25 t ha^{-1} along with 75 kg ha^{-1} P_2O_5 through various inorganic fertilizers recorded the highest available P in the soil of a rice based cropping system.

In a long-term experiment conducted at Panthnagar, India, Rathore *et al.* (2004) revealed that the application of recommended dose of NPK along with 10 t ha^{-1} of FYM in the soils under rice- wheat system increased the available P over the inorganic fertilizer alone.

Mishra *et al.* (2008) also reported that phosphorus content of soil increased due to the application of organic sources of nitrogen. Higher amount of available P observed in biofertilizers treated plots might be attributed to solubilizing effect of PSB on mineral and unavailable form of P. Solubilization of P associated with pH drop in PSB treated plots indicated that organic acid production by PSB played a significant role in acidification facilitating P solubilization (Hwangbo *et al.* 2003).

The combined use of organic and inorganic fertilizers has been reported not only to meet the nutrients need of the crop but also has been fund to sustain large scale productivity goals (Yadav and Meena 2014). Sabina Ahmmmed et al. (2014) showed that combined application of 120 kg N + 39.3 kg P + 100 kg K ha along with 10 t FYM ha^{-1} resulted in the build up of organic carbon, available N and P in soil.

Howida Elhabet (2018) reported that 4.76 t of composted rice straw + 110 kg N ha^{-1} recorded the highest amounts of nutrients (NH_4^+ and P) under flooded condition.The soil available N, P_2O_5, K_2O content of soil significantly increased with the application of Zn enriched FYM.

To study the efficacy of bioresources on the soil phosphorus availability, the treatments involving T_1 - Absolute control, T_2 - 100% NPK, T_3 - 75% P + 25% P through enriched FYM + NK, T_4 - 75% P + *Phosphobacterium* + NK, T_5 - 75% P + 25% P through Casuarina needles + NK, T_6 - 50% P + 50% P through enriched FYM + NK, T_7 - 50% P + *Phosphobacterium* + NK, T_8 - 50% P + 50% P through Casuarina needles + NK, T_9 - 25% P + 75% P through enriched FYM + NK, T_{10} - 25% P + *Phosphobacterium* + NK, T_{11} - 25% P + 75% P through Casuarina needles + NK were studied using Completely Randomised Design with three replications. The results obtained with respect to phosphorus availability is discussed with suitable literatures.

Incubation Experiment

Olsen-P

The increased Olsen's P content in the coastal salt affected soil due to the application of various bio-resources was well evidenced in both the incubation experiments (Table 1). The experimental soil exhibited low status of available phosphorus. Bio-resources application played in a greater way for the improvement of available P status of the soil. The increased availability of phosphorus due to the application of bio-resources was in the range of 8.13 to 8.69 mg kg^{-1} in the incubation experiment I (nitrogen) and 5.95 to 8.05 mg kg^{-1} in the incubation experiment II (phosphorus) as compared to 6.02 and 4.55 mg kg^{-1} in control of first and second incubation experiments respectively. The substitution of 25 % N with *Sesbania rostrata* (8.13 mg kg^{-1} at 90 DAI) in the N incubation experiment and 25 % P with enriched FYM (8.05 mg kg^{-1} at 90 DAI) in the P incubation experiment had greater significance in increasing the availability of P. The inoculation of *Phosphobacterium* also rated best in increasing the P availability. Application of composted coirpith, *Casuarina* needles and *Azospirillum* also had significant influence in increasing the P availability as compared to control.

Generally, the P content of the added inputs is the most important factor for phosphorus mineralization in the soil (Hundal and Thind, 1993). In the incubation studies, enriched FYM and *Casuarina* needles were used as the bio-resource source for supplementing P nutrition. The enriched FYM and *Casuarina* needles contained 0.86 per cent and 0.28 per cent of P respectively. In the N incubation experiment also the applied green manure contained 0.48 per cent and the composted coirpith contained 0.15 per cent of P. During the process of decomposition and mineralization, the P released from the added bio-resources greatly contributed to it's availability and this might explain the superiority of enriched FYM and green manure in increasing the phosphorus availability in the soil. The application of *Phosphobacterium* increased the availability of P in the soil. The phosphorus solubilizing bacteria have the capacity to produce acetic, citric and lactic acids (Hayman, 1975) with which the insoluble phosphate get solubilized and the P availability becomes higher.

The submerged condition of the incubated soil further stabilized the P mineralization by the way of (i) its solubilizing effect on the decomposed product and (ii) proliferating phytase producing microorganisms (Vig *et al.*, 1997). Prasad and Singharia (1989) also reported the increased P availability in soil with the addition of enriched organic manures.

Further, production of organic acids released during the process of decomposition might have solubilized the insoluble form of P to soluble fraction (Singh Parihar *et al.*, 2003) resulted in increased P availability as evidenced in the incubation studies.

Table 1. Effect of bio-resource treatments on the available Phosphorus content (mg kg^{-1}) of soil in incubation experiments

	Incubation Experiment 1			**Incubation Experiment 2**		
Treatments	**30 DAI**	**60 DAI**	**90 DAI**	**30 DAI**	**60 DAI**	**90 DAI**
T_1	5.92	5.96	6.02	4.32	4.43	4.55
T_2	6.49	6.96	7.07	6.48	6.62	7.45
T_3	7.65	7.95	8.69	7.60	7.87	8.05
T_4	7.22	7.64	8.29	7.50	7.68	7.80
T_5	7.42	7.68	8.44	7.22	7.45	7.60
T_6	6.87	7.35	8.20	6.68	6.87	7.03
T_7	6.50	7.32	8.32	6.42	6.72	6.77
T_8	6.71	7.22	8.38	5.98	6.23	6.42
T_9	6.98	7.20	8.36	5.92	6.10	6.27
T_{10}	6.58	7.32	8.18	5.60	5.63	5.95
T_{11}	6.54	7.15	8.13	5.35	5.54	5.68
SEd	0.11	0.14	0.12	0.16	0.21	0.15
CD (p=0.05)	0.21	0.27	0.24	0.32	0.41	0.29

Pot and field Experiment

Olsen- P

The experimental soil showed low available phosphorus status (8.0 kg ha^{-1}). The P in coastal saline soils is variable depending on the nature of parent material, degree of salinity and physico chemical properties of the soil. Ryden and Syers (1975), Singh *et al.* (1990) and Singh *et al.* (1992) reported the reduced availability of P in coastal saline soils.

In the present investigation, a profound influence of all the applied bio-resources in increasing the Olson's P in the coastal salt affected soil was well documented (Table 2&3). Among the bio-resource treatments, the addition of 75% NP + 25 % N supplied through green manure + 25% P through single super phosphate enriched FYM significantly increased the Olsen- P in the soil. The application of *Phosphobacterium* and green manure also recorded the higher P availability over the control treatment. In the pot experiment, the highest P concentration of 5.64 mg kg^{-1} at the harvest stage was recorded with the application of green manure + enriched FYM. The results of field experiments also indicated the usefulness of this treatment in increasing the P availability (10.40 and 11.00 kg ha^{-1} at the end of first and second crop respectively). This result was in accordance with the findings of Bandyopadhyay and Rao (2001) who stated that the use of bulky organic manures and green manures are very beneficial for P management of saline soils as they improve the soil physical condition and increases the availability of soil phosphates.

Manures are always much more effective than soluble inorganic P fertilizers in increasing the available P in the soil (Broadbent, 1986). The application of enriched FYM resulted in the production of organic acids during the process of decomposition, which solubilized the insoluble form of P to soluble fraction (Singh Parihar *et al.*, 2003; Malarkodi and Singaram, 2004). The green manure played a significant role in mobilizing the native soil P and reduction in P fixation during its decomposition and more phosphates placed in the labile pool (Rathore *et al.*, 2004). Further, the green manure act as the effective sink for Ca, which released during the process of decomposition. The released Ca might be chelated with the native P and enhanced its availability (Selvi *et al.*, 2003a).

The application of *Phosphobacterium* increased the availability of P in the soil. The phosphorus solubilizing bacteria produced acetic, citric and lactic acids (Hayman, 1975). With the aid of these organic acids, the micro flora released the P from the soil matrix (Arden-Clarke and Hodges, 1988), which resulted in the production of orthophosphate ions, the form of P most commonly taken up by plants (Hayman, 1975). Also the high pH of the soil stimulated the cellulolytic microorganisms and the decomposition of applied bio-resources in the soil gets hastened (Szegi, 1988). The metabolites of the decay products or the cellulolytic microorganisms themselves may enforce P release, because of the formation of organic acids, which dissolved phosphates and/or hindered P fixation (Tan, 1986).

Selvi *et al.* (2003a), Ravichandran *et al.* (2003), Rathore *et al.* (2004) and Laxminarayana (2005) were also reported earlier the increased P availability due to the application of various bio-resources.

Table 2. Effect of bio-resource treatments on the available Phosphorus (mg kg^{-1}) content of the soil in the pot experiment

Available P		
AT	**PI**	**Har**
4.83	4.77	4.73
5.07	5.01	4.96
4.94	4.89	4.85
5.26	5.19	5.12
5.14	5.08	5.03
5.18	5.14	5.09
5.76	5.69	5.64
5.43	5.37	5.31
5.17	5.11	5.08
0.14	0.18	0.13
0.28	0.36	0.25

T_1 – Control (100% NPK), T_2 - 75% N + 25% N through *Sesbania rostrata* + PK, T_3 - 75% N + *Azospirillum* + PK, T_4 - 75% P + 25% P through enriched Farm yard manure + NK, T_5 - 75% P + *Phosphobacterium* + NK, T_6 - 75% N + 25% N through *Sesbania rostrata* + 75% P + 25% P through enriched Farm yard manure + K, T_7 - 75% N + 25% N through *Sesbania rostrata* + 75% P + *Phosphobacterium* + K, T_8 - 75% N + *Azospirillum* + 75% P + 25% P through enriched Farm yard manure + K, T_9 - 75% N + *Azospirillum* + 75% P + *Phosphobacterium* + K

Table 3. Effect of bio-resource treatments on the available Phosphorus (kg ha^{-1}) content of the soil in the field experiments

Treatments	Field experiment 1			Field experiment 2		
	AT	PI	Har	AT	PI	Har
T_1	10.55	8.85	7.55	10.35	8.65	7.05
T_2	15.40	13.20	10.40	16.45	13.20	11.00
T_3	13.90	10.58	9.23	14.00	12.65	10.65
T_4	13.10	9.80	8.55	13.25	12.00	8.60
T_5	12.80	9.10	8.18	11.90	9.90	8.23
SEd	0.39	0.35	0.08	0.34	0.32	0.37
CD (p=0.05%)	0.78	0.71	0.17	0.68	0.63	0.74

T_1- Control (100% NPK), T_2- 75% NP + 25% N through *Sesbania rostrata* + 25% P through Super Phosphate enriched FYM + K, T_3- 75% NP + 25% N through *Sesbania rostrata* + *Phosphobacterium* + K, T_4- 75% NP + *Azospirillum* + 25% P through Super Phosphate enriched FYM + K, T_5- 75% NP + *Azospirillum* + *Phosphobacterium* + K

5

Bioresources and Soil Potassium

Potassium availability in most of the coastal salt affected soils of India is rich except when the soils are low in clay content (Bandyopadhyay *et al.*, 1985). Plants grown under high salinity may show K deficiency due to the antagonistic effect of Na on K absorption and / or disturbed Na / K ratio. Under such conditions application of K is also known to reduce the adverse effect of salinity and sodicity on crop growth (Qadar, 1995). Singh *et al.* (1980) reported that the addition of FYM increased the K concentration in the soil. Swarup (1987) reported that the incorporation of green manure increased the available N, P and K status of the soil. Available P and K remarkably increased to a tune of 30 and 44 per cent respectively by the application of coir waste (Santhi *et al.*, 1991).

Marked increase in soil *Casuarina* needles along with gypsum amendment in coastal saline soil.solution K with the addition of *Sesbania* and dhaincha was reported by Yadvinder Singh *et al.* (1992) and Mahapatra and Jee (1993). Mishra and Sharma (1997) reported a marked increase in the available N, P and K status of soils of rice based cropping systems with the applications of FYM + 100 per cent RDF.

Yadav *et al.* (2000) found that the application of 50 percent RDF and remaining nutrients through FYM or green manure (*Sesbania sp.*) increased the K content in the soil of rice- wheat system. Yaduvanshi (2001) reported a considerable increase in available K level of a reclaimed sodic soil due to the application of FYM @ 10 Mg ha^{-1}. Selvi *et al.* (2003b) revealed that the available N and K in the soil increased with the addition of 50 per cent N and K along with green manuring (*Sesbania aculeata*). Singaravel *et al.* (2001) stated an increase in the NPK status of rice sodic soil with the application of *Casuarina* needles along with gypsum. Rathore *et al.* (2004) reported an increase in the soil K level due to the addition of FYM along with recommended chemical fertilizers to the soil. Sabina Ahmmmed *et al.* (2014) showed that the highest amount of available K was found in the treatment receiving 120 kg N, 26.2 kg P ha^{-1} and 100 kg K ha^{-1} along with 10 t FYM ha^{-1}.

Sai Ram *et al.* (2020) revealed that with 75% RDN+25% N through vermicompost and the availability soil potassium increased to a level of K 144.0 kg/ha in postharvest soil.

The results obtained from the incubation, pot and field experiments with respect to K availability is discussed in line with available literature.

Incubation Experiment

NH_4OAc- K

In both the incubation experiments, the entire dose of K was applied as inorganic form. Application of various bio-resources showed an improvement in the K content of the soil (Table 1).

In the N incubation experiment, the highest K availability was observed with the application of green manure, which registered 195.00 mg kg^{-1} at 90 DAI. The other bio-resources viz. composted coirpith and *Azospirillum* along with 75 per cent N also improved the K status of the soil. In the second incubation experiment, the application of enriched FYM improved the K content significantly (190.67 mg kg^{-1}) over other treatments. The addition of *Casuarina* needles and *Phosphobacterium* also contributed towards the improvement in K content of the soil as compared to control.

The improved N availability observed in the present study with the various bio-resource treatments might bring about more K into soil solution by the way of exchange reaction. The NH_4^+ ion having greater replacing power and preferential absorption characteristics over K^+ would have lead to constant release of exchangeable K from the soil lattice into the soil solution and contributed to the increased NH_4OAc- K content of the soil. The differences among the bio-resources treatments in increasing the K availability could be ascribed to the variation in the amounts of total K added through bio-resources, their rate and amount of mineralization (Yadvinder Singh *et al.*, 2005). Earlier reports of Bulluck III *et al.* (2002) and Kulvinder Kaur *et al.* (2005) lend support to the present findings, that K concentration in soil amended with organic wastes increased over soils applied with chemical fertilizers.

In both the incubation experiments, the higher availability of nutrients were recorded with 75 per cent N and P through fertilizers and remaining substituted through various bio-resources than 50 or 25 per cent inorganic fertilizers and remaining substituted through various bio-resources. This might be ascribed to the restricted availability of nutrients due to immobilization when the bio-resources are applied in relatively larger quantities (Mukherjee and Agarwal, 1951; Maharudrappa *et al.*, 2000).

Table 1. Effect of bio-resource treatments on the NH_4OAC-K (mg kg^{-1}) content of the soil in the incubation experiment

Treatments	N Incubation experiment			P Incubation Experiment		
	NH_4OAC-K (mg kg^{-1})			NH_4OAC-K (mg kg^{-1})		
	30 DAI	60 DAI	90DAI	30 DAI	60 DAI	90 DAI
T_1	148.33	148.80	149.50	120.83	125.08	126.17
T_2	153.60	155.25	159.83	141.67	146.50	150.33
T_3	187.50	192.50	195.00	175.67	185.33	190.67
T_4	179.00	181.66	183.00	162.33	164.33	168.67
T_5	180.00	183.33	186.50	172.33	173.83	177.50
T_6	168.52	170.83	173.67	160.67	166.67	169.83
T_7	155.63	158.58	161.83	154.50	159.67	162.17
T_8	158.58	163.70	166.50	162.00	164.65	167.83
T_9	161.70	164.66	165.17	160.50	163.64	166.33
T_{10}	153.83	156.66	158.50	152.33	156.00	159.33
T_{11}	157.66	161.16	164.66	158.50	160.67	162.67
SEd	3.89	4.18	4.56	4.71	4.44	4.35
CD (p=0.05)	7.78	8.36	9.12	9.42	8.87	8.69

Pot and field Experiment

Generally the coastal salt affected soils contain adequate amount of available potassium (Bandyopadhyay *et al.*, 1985). Application of K fertilizers have the benefit of reducing the adverse effect of salinity and sodicity on the crop growth by the way of adjusting Na / K ratio in these soils (Mengal and Kirkby, 1982).

The increase in the K concentration of the soil due to the application of various bio-resources was well evidenced in the present investigation (Table 2&3). Among the bio-resource treatments, the application of 75 % NP + green manure + enriched FYM recorded the highest K concentration of the soil in pot and field experiments. This treatment recorded 144.50 mg kg^{-1} in the pot experiment and 277.56 kg ha^{-1} at the end of second field crop respectively at the harvest stage. Several authors reported such application (inorganic and organic) in the build up of K in the soil (Chaudhary *et al.*, 1981; More, 1994; Prasad *et al.*, 2002; Rathore *et al.*, 2004).

The increased K availability was due to the addition of K being added through green manure and FYM. The minimal loss of K through leaching as the result of fixation of K ions on the cation exchange sites might explain the increased K availability. Selvi *et al.* (2003b) also had similar opinion regarding the K availability in the soil with the applied organics.

The increased K concentration due to the addition of biofertilizers along with green manure and FYM could be attributed to the release of organic and inorganic acids that might have solubilized K from primary minerals of the soil.

Table 2. Effect of bio-resource treatments on the NH_4OAC-K ($mg\ kg^{-1}$) content of the soil in the pot experiment

Treatments	NH_4OAC-K		
	AT	PI	Har
T_1	128.56	123.70	120.56
T_2	145.72	141.00	137.81
T_3	133.56	128.56	124.44
T_4	140.76	136.28	131.76
T_5	134.72	129.59	125.25
T_6	152.72	148.62	144.50
T_7	149.27	145.77	140.27
T_8	143.55	139.43	135.64
T_9	137.33	132.65	128.79
SEd	2.16	2.58	2.13
CD (p=0.05%)	4.32	5.16	4.26

T_1 - Control (100% NPK), T_2 - 75% N + 25% N through *Sesbania rostrata* +PK, T_3 - 75% N + *Azospirillum* + PK, T_4 - 75% P + 25% P through enriched Farm yard manure + NK, T_5 - 75% P + *Phosphobacterium* + NK, T_6 - 75% N + 25% N through *Sesbania rostrata* + 75% P + 25% P through enriched Farm yard manure + K, T_7 - 75% N + 25% N through *Sesbania rostrata* + 75% P + *Phosphobacterium* + K, T_8 - 75% N + *Azospirillum* + 75% P + 25% P through enriched Farm yard manure + K, T_9 - 75% N + *Azospirillum* + 75% P + *Phosphobacterium* + K

Table 3. Effect of bio-resources on the NH_4OAC-K content of the soil in the field experiments

	Field Experiment 1			Field Experiment 2		
Treatments	NH_4OAC-K			NH_4OAC-K		
	AT	PI	Har	AT	PI	Har
T_1	275.72	265.55	255.82	272.80	261.55	251.70
T_2	307.10	286.72	276.43	294.17	283.10	277.56
T_3	280.37	277.81	271.55	295.55	281.47	272.72
T_4	291.25	270.47	264.22	271.56	277.10	265.58
T_5	287.58	269.52	259.26	270.18	263.56	260.17
SEd	5.67	8.25	9.29	5.61	5.11	3.86
CD (p=0.05%)	10.18	16.59	18.67	11.22	10.21	7.72

T_1- Control (100% NPK), T_2- 75% NP + 25% N through *Sesbania rostrata* + 25% P through Super Phosphate enriched FYM + K, T_3- 75% NP + 25% N through *Sesbania rostrata* + *Phosphobacterium* + K, T_4- 75% NP + *Azospirillum* + 25% P through Super Phosphate enriched FYM + K, T_5- 75% NP + *Azospirillum* + *Phosphobacterium* + K

6

Bioresources and Secondary and Micronutrients

Secondary nutrients

Secondary nutrients includes Calcium, magnesium, and sulfur. Crop plants requires secondary nutrients in moderate quantity for crop growth and yield. Their requirement by plant is less than the macro nutrients. Sulphur and Magnesium requirement of crops are very similar to the Phosphorus needs. Calcium is required in higher amount than Phosphorus. Calcium and magnesium are mostly deficient in acid and alkali soils. In India S deficiencies are found in almost 40 per cent of the cultivated soil. Under intensive cropping system, the use of higher amount of nitrogenous fertilizer without secondary and micro nutrients and lesser use of organic manures are the major reason for the secondary and micro nutrient deficiency in soil.

Mathur *et al.* (1983) demonstrated that application of 2 to 4 q ha^{-1} at the time of sowing has increased the yield by 48 per cent. Gupta and Abrol (1990) demonstrated the significant effect of calcium in the remediation of sodic soil. Magnesium forms the essential constituent in chlorophyll. It combines with ATP, which participates in many metabolic reactions thus it is highly essential for growth and development of plants. Sulphur is a constituents in methionine and cysteine aminoacids which forms proteins (Thompson *et al.*, 1966). Increased secondary nutrient availability and its uptake by plant will improve higher crop productivity.

Improvement in secondary nutrient Ca to tune of 25 per cent with pressmud application was evidenced by Shivakant and Rajkumar (1992). Lavanya and Manickam (1991) recorded significant improvement in the availability of calcium and magnesium with the application of compost and biogas slurry. In sodic soil, Anand Swarup (1992) evidenced an increase in the exchangeable Ca^{++} and Mg^{++} throughout the rice-growing period with the addition of *Sesbania aculeata*. Continued application of manure increased the calcium and magnesium in soil (Patiram and Singh, 1993). Application of compost and rice straw incorporation increased the available Ca and Mg content in soil (Nahar *et al.*, 1995).

Practicing long-term Integrated Nutrient Management with the incorporation of FYM along with fertilizers in soil increased the fertility status and leads to the increased content of secondary nutrient Ca and Mg (Bellaki *et al.,* 1998). Organic matter incorporation along with gypsum increased the exchangeable calcium in the soil (Vance and McKenzie, 1998).

Application of enriched compost increased the magnesium content in the soil (Srikanth *et al.,* 2000). The FYM with its high adsorptive capacity, retain more Ca^{2+} and Mg^{2+} ions, prevent leaching and make them available to crop plants. Selvi *et al.* (2003) also opined that that the exchangeable Ca and Mg increased with the application of recommended dose of NPK fertilizer with Farm Yard Manure.

Increased availability of sulphur content up to 32 per cent in soil with the application of green manures was reported by Mythili *et al.* (2001). Reddy and Raja Reddy (2001) also observed that the S content was higher in soils treated with organic manures, when compared to control. An increase in available sulphur in soil was recoded with the incorporation of 5 t ha^{-1} of FYM & 6 t ha^{-1} of green manure (*Parthenium sp.*) together with RDF (Muneshwar Singh *et al.*, 2001).

S release followed an initial rapid rate upto 10 weeks and a relatively slow release between 10 and 16 weeks. The amount of S mineralized was higher in organic manures (61-67%) as compared to green manures (50-56%) and the mineralization pattern was identical in vertisol and inceptisol. Sriramachandrasekaran *et al.* (2004) noted an increase in the S content of the soil with the addition of sulphitation pressmud @10 t ha^{-1} + FYM @ 12.5 t ha^{-1} + RDF in a sulphur deficient rice soil.

Application of NPK fertilizer along with compost and lime increased the exchangeable Ca in soil (Takeda *et al.,* 2005). Babu *et al.* (2007) also proved that Farm Yard Manure addition significantly improved the exchangeable Ca and Mg in soil.

Pot experiment including treatments as given in Table 1 and field experiments including treatments as given in Table 2 were evaluated to study the influence of bioresources on the Secondary nutrient availability for rice. The experimental results were explained with available scientific evidences accumulated in the past.

Sulphur

In the present study, the application of various bio-resources significantly increased the S content of the soil (Table 1). The application of green manure with enriched FYM and green manure + *Phosphobacterium* either alone or in combination with other bio-resources proved more efficient in increasing the S availability. The soil incorporation of green manure and enriched FYM as the sources for 25% N and P respectively recorded the highest S content of the soil in the pot experiment and its validity was proved in the field experiments as well. This treatment recorded the higher S concentration of 15.51 mg kg^{-1} in the pot experiment, 15.76 mg kg^{-1} in the first year and 15.83 mg kg^{-1} in the second year field experiment.

The increased availability of S in the soil on incorporation of green manure and or enriched FYM along with NPK fertilizers was due to the process of mineralization of organic S on decomposition (Muneshwar Singh *et al.*, 2001). Sulphur occurs in plants in compounds such as methionine, cystine, cystein, thiamine, coenzyme A, biotine, thioctic acid, thiocyanates, glutatione, alkaloids as well as the intermediates associated with the synthesis of these and other plant substances (Thompson *et al.*, 1966). The steady release of S from the added bio-resources might explain the microbial decomposition of these S containing materials, when incorporated into the soil.

The increased substrate availability by the way of increased microbial decomposition of applied bio-resources as evidenced in the present study might have influenced the S mineralization favorably as suggested by Frenzy (1986). Earlier workers Muneshwar Singh *et al.* (2001) and Sriramachandrasekaran *et al.* (2004) also reported an increased S concentration in the soil due to the application of various bio-resources.

Table 1. Effect of bio-resource treatments on the secondary nutrients availability in the pot experiment

Treatments	S (mg kg^{-1})			Ca (C mol kg^{-1})			Mg (C mol kg^{-1})		
	AT	PI	Har	AT	PI	Har	AT	PI	Har
T_1	13.45	12.08	11.97	3.88	3.67	2.32	3.33	3.16	2.89
T_2	18.67	17.43	15.37	4.92	4.76	3.93	4.22	3.27	3.09
T_3	15.49	14.12	13.02	3.89	2.69	2.49	3.34	3.19	2.95
T_4	17.56	16.32	14.25	4.56	4.21	3.52	3.38	3.22	3.02
T_5	15.50	14.13	13.00	3.89	3.70	3.48	3.35	3.19	2.94
T_6	18.74	17.59	15.51	5.98	4.82	4.57	4.48	3.32	3.37
T_7	18.69	17.47	15.39	5.94	4.77	4.36	4.40	3.28	3.21
T_8	17.60	16.35	14.28	4.72	4.38	3.66	3.44	3.25	3.05
T_9	15.56	14.15	13.04	3.90	3.71	3.50	3.35	3.20	2.95
SEd	0.15	0.16	0.10	0.02	0.02	0.02	0.01	0.02	0.02
CD (p=0.05%)	0.30	0.32	0.21	0.03	0.04	0.03	0.02	0.03	0.03

T_1 - Control (100% NPK), T_2 - 75% N + 25% N through *Sesbania rostrata* +PK, T_3 - 75% N + *Azospirillum* + PK, T_4 - 75% P + 25% P through enriched Farm yard manure + NK, T_5 - 75% P + *Phosphobacterium* + NK, T_6 - 75% N + 25% N through *Sesbania rostrata* + 75% P + 25% P through enriched Farm yard manure + K, T_7 - 75% N + 25% N through *Sesbania rostrata* + 75% P + *Phosphobacterium* + K, T_8 - 75% N + *Azospirillum* + 75% P + 25% P through enriched Farm yard manure + K, T_9 - 75% N + *Azospirillum* + 75% P + *Phosphobacterium* + K

Calcium and magnesium

The application of bio-resources significantly increased the exchangeable Ca and Mg in comparison to chemical fertilizers. Among the bio-resource treatments, the highest Ca content of 4.57 C mol P $^{(+)}$ kg^{-1} in the pot experiment and 4.99 in field experiment one and 5.04 C mol P $^{(+)}$ kg^{-1} in the field experiments two was recorded with the application of *Sesbania rostrata* + enriched FYM + 75 per cent NP (Table 2&3). This treatment also recorded the highest Mg concentration in pot (3.37 C mol P $^{(+)}$ kg^{-1}) and field experiments.

Submergence of sodic soil leads to the accumulation of CO_2, which solubilizes the native $CaCO_3$ and Ca thus released in turn cause its increased availability. The green manure and enriched FYM generated more CO_2 in the rice soil and the component Ca and Mg of the bio-resources released during the decomposition made significant contribution in increasing the secondary nutrient Ca and Mg (Anand Swarup, 1992). Due to the microbial activity of rice roots, CO_2 and other secretions liberated also contributed towards the increased exchangeable Ca and Mg.

The mechanism of high adsorptive capacity of FYM and green manure helped to hold more calcium and magnesium that would otherwise be leached thus making them available in the soil (Prasad and Singh, 1981; Selvi *et al.*, 2003b). Srikanth *et al.* (2000) also reported an increased secondary nutrient availability with the application of various bio-resources in the soil.

Table 2. Effect of bio-resource treatments on the secondary nutrients availability in the field experiment- I

	S (mg kg^{-1})			Ca (cmol kg^{-1})			Mg (cmol kg^{-1})		
Treatments	**AT**	**PI**	**Har**	**AT**	**PI**	**Har**	**AT**	**PI**	**Har**
T_1	15.35	14.05	13.25	3.92	3.81	3.70	3.47	3.42	3.35
T_2	17.05	15.98	15.76	5.07	5.02	4.99	4.05	4.00	3.96
T_3	16.87	15.63	15.50	5.01	4.95	4.92	4.01	3.97	3.90
T_4	16.52	15.06	15.11	4.97	4.90	4.86	3.97	3.91	3.84
T_5	14.88	15.22	13.65	3.94	3.85	3.74	3.51	3.43	3.37
SEd	0.40	0.41	0.37	0.03	0.03	0.03	0.03	0.02	0.03
CD (p=0.05%)	0.79	0.82	0.73	0.05	0.05	0.06	0.04	0.04	0.05

T_1- Control (100% NPK), T_2- 75% NP + 25% N through *Sesbania rostrata* + 25% P through Super Phosphate enriched FYM + K, T_3- 75% NP + 25% N through *Sesbania rostrata* + *Phosphobacterium* + K, T_4- 75% NP + *Azospirillum* + 25% P through Super Phosphate enriched FYM + K, T_5- 75% NP + *Azospirillum* + *Phosphobacterium* + K

Table 3. Effect of bio-resource treatment on the secondary nutrients availability in the field experiment-II

	S (mg kg^{-1})			Ca (cmol (P^{+})kg^{-1})			Mg (cmol (P^{+}) kg^{-1})		
Treatments	AT	PI	Har	AT	PI	Har	AT	PI	Har
T_1	14.52	13.58	12.65	3.72	3.65	3.56	3.37	3.22	3.15
T_2	17.60	16.37	15.83	5.20	5.11	5.04	4.21	4.05	3.99
T_3	17.92	16.27	15.44	5.09	5.00	4.96	4.16	3.98	3.92
T_4	17.07	16.13	15.26	5.01	4.93	4.92	4.05	3.95	3.87
T_5	15.92	15.45	13.95	3.89	3.81	3.77	3.92	3.57	3.40
SEd	0.13	0.10	0.11	0.16	0.12	0.11	0.07	0.08	0.06
CD (p=0.05%)	0.26	0.19	0.21	0.32	0.23	0.21	0.13	0.16	0.12

Micronutrients

Various studies have shown that application of high dose of inorganic fertilizers alone can exacerbate micronutrient deficiency by reducing their availability (Prasad *et al.*, 2000). The saline soils, particularly in the coastal regions, are well supplied with micronutrients except zinc deficiency. Results of the various experiments conducted in different coastal regions of India revealed that micronutrients availability increased with organic matter incorporation (Bandyopadhyay and Rao, 2001).

Nagendra Rao (1985) reported that the application of farmyard manure increased significantly the content of DTPA extractable Fe, Mn and Zn in the soil.

Kher (1993) showed that application of FYM @ 10 t ha^{-1} along with 100% NPK increased the micronutrient availability in silt loam soil. The mineralization of organically bound forms of Zn in the FYM and formation of organic chelates of higher stability, as Zn reduced the fixation and precipitation. The increased Cu availability was attributed to the formation of Cu-humates complexes in soil.

Anand Swarup (1988) also evidenced that the application of organics viz., *Casuarina equisetifolica* and *Acacia milotica* significantly increased the concentration of Fe and Mn in rice grown in sodic soil. Through a long-term field experiment in sodic soil, Kumar *et al.* (1995) showed that rate of depletion of DTPA extractable micronutrients was higher in inorganic fertilizers compared to the treatments with organic manures like FYM and *Sesbania* green manure. Organic compounds released during decomposition of organic sources increased the availability of micronutrients by chelation action.

Long-term effect of continuous rotational cropping and fertilization on soil properties of an alfisol showed that FYM + 100 per cent NPK increased the available micronutrients significantly (Bharadwaj and Omenwar, 1994).

Prasad *et al.* (1996) found a considerable increase in available Fe and Zn content with the application of poultry manure in calcareous soil. Complexing properties of poultry manure may prevent precipitation and fixation of Fe and Zn and keep them in slowly available form.

Mathur (1997) also evidenced that FYM incorporation increased available Zn and Mn status of soil. The mineralization of FYM and consequent release of micronutrients explains the reason for increased availability. Similar observation was also recorded by Santhi et al.(1998) with FYM @ 10 t ha^{-1} along with 100 per cent NPK in increasing micronutrient availability.

Research work at Central Soil Salinity Research Institute, Karnal also revealed that available Zn and Mn were higher in organic and inorganic source of nutrients applied in combination than inorganic source alone. Incorporation of FYM @ 10 t ha^{-1} along with inorganic fertilizer for a longer period increased the DTPA- Zn and Mn respectively from 0.80 to 0.92 ppm and 3.7 to 4.2 ppm (Swarup and Yaduvanshi, 2000).

Nayar and Chhibba (2000) reported that application of green manure or farmyard manure significantly increased the availability of all the micronutrients. The increase was more significant with 50 per cent NPK supplied through FYM & green manure in combination with chemical fertilizers. FYM along with 100 per cent NPK proved effective in increasing the available micronutrients (Selvi *et al.*, 2003b).

Latha *et al.*(2001) reported that zinc enriched organic manures enhanced the concentration of all forms of Zinc except crystalline iron oxide bound zinc. Kotha Sammy Reddy (2002) found an increased cumulative S mineralization in soils amended with organic manures (FYM or wheat straw) and green manures (*Leucaena leucocephala* or *Gliricidia sepium*).

FYM applied plots showed an increased content of micronutrients viz., Fe, Mn, Zn and Cu with FYM application after thirty five years of cropping (Szalai *et al.*, 2002). Sindhu and Sharma (2010) reported that availability of micronutrients increased with the increase in organic carbon content and NPK + green manuring increased the Mn availability than 100% NPK alone (Verma *et al.*, 2005).

Available Zn, Cu, Mn, and Fe contents increased with Sesbania green-manure along with NPK than control (Yadav and Kumar, 2009).

In a study conducted at IARI, involving treatments including control, 100% N, 100% NP, 100% NPK, 100% NPK+FYM and 100% NPK+Zn, highest available Zn increased with 100% NPK+FYM as compared to control (Behera and Singh, 2009).

Myint *et al.* (2010) reported that the use of organic matter increases various types of nutrients needed by plants, including micronutrients, improve soil physical and chemical properties and microbial activity. increased The application of green manures and FYM as compared to 100 percent NPK was reported to increase DTPA-extractable Zn, Cu, Fe and Mn content in soil (Kumar and Singh, 2010).

Complexing of Zn with dissolved organic compounds increases Zn solubility and mobility (Houben and Sonnet, 2012). To maximize crop yields, the use of organic manures combined with inorganic fertilizers is necessary to make the soil rich in organic matter in the form of readily available nutrients (Ramalakshmi *et al.*, 2012).

Crop residue incorporation was shown to improve micronutrient availability in soil. Rathood *et al.* (2013) evidenced an increase in DTPA- Fe, Zn, and Cu with the addition of farmyard manure and vermin compost addition.

Organic matter is a key component in soil that accumulates from the breakdown of animal and plant residues. Humic substances including humic acid and fulvic acids posses a have variety of functional groups like OH, COOH, SH that have a strong affinity for metals such as Zn, and increases the dissolution and movement over wide pH range (Hafeez *et al.,* 2013). An increase in availability up to 23 and 31 per cent of calcium and magnesium respectively in manured over non manured plots was reported by Iqbal *et al.* (2014).

An increase in DTPA-Zn with the mineralization of organically bound forms of Zn in the from FYM treatment was reported by Shanbavi *et al.* (2018). Application of NPK + FYM significantly enhanced the available of zinc, iron, manganese, and copper over control (Bora *et al.*, 2018). Manoj *et al.* (2019) evidenced that balanced application of inorganic fertilizer with FYM in rice, while without FYM in wheat significantly improved the DTPA-extractable Zn, Fe, Mn and Cu after rice and wheat crops.

Field experiments were conducted in randomised block design (RBD) with four replications involving the following five treatments: T_1- Control (100% NPK), T_2- 75% NP + 25% N through *Sesbania rostrata* + 25% P through Super Phosphate enriched FYM + K, T_3- 75% NP + 25% N through *Sesbania rostrata* + *Phosphobacterium* + K, T_4- 75% NP + *Azospirillum* + 25% P through Super Phosphate enriched FYM + K, T_5 - 75% NP + *Azospirillum* + *Phosphobacterium* + K. The results obtained with respect to DTPA extractable micronutrients are discussed with available scientific evidences.

In general, most of the saline soils, including the coastal region are well supplied with micronutrients with exception of Zn. The deficiency of micronutrients are induced in saline rice growing soils due to the reduction of sulphate salts to sulphides and subsequent precipitation of the micronutrients (Maji and Bandyopadhyay, 1990; Subba Rao *et al.*, 1994).

In the present investigation, the analysis of micronutrient status of the experimental soils at all the critical stages of rice growth clearly indicated the efficiency of the bio-resources in increasing the DTPA- extractable micronutrients (Table 4&5). Of the various bio-resources, the application of 75 % NP + *Sesbania rostrata* + enriched FYM registered the highest availability of Fe (9.28 and 11.12 mg kg^{-1}), Mn (8.18 and 9.22 mg kg^{-1}), Cu (0.75 and 0.79 mg kg^{-1}) and Zn (0.77 and 0.83 mg kg^{-1}) at the harvest of first and second year of rice cultivation respectively. The other bio-resource treatments namely *Sesbania rostrata* + *Phosphobacterium*, *Azospirillum* + enriched FYM, *Azospirillum* + *Phosphobacterium* also significantly improved the micronutrient status of the soil in comparison to control.

Application of green manure or farmyard manure is known to increase concentrations of Fe, Mn, Cu and Zn in rice based cropping system (Nayar and Chibba, 2000). Under anaerobic conditions, especially when organic manures are incorporated in the paddy soil leads to strongly reduced environment where molecular oxygen is depleted and Fe (Fe_2O_3, $Fe_3O_4.nH_2O$ and $Fe(OH)_3.nH_2O$) and Mn (Mn_2O_3 and Mn_3O_4) are chemically reduced to Fe^{2+} and Mn^{2+} (Ponnamperuma, 1972; Yu, 1985; Yang *et al.*, 2002). As a result, the concentrations of reduced substances are greatly promoted in the paddy soils. The decrease in pH and ESP of sodic soils upon submergence and application of organic matter also facilitated this reduction process (Anand Swarup, 1992).

The greater availability of Zn could be due to the decrease of soil pH contributed through added bio-resources. The negative correlation of Zn with pH was reported by many authors (Bhumbla, 1972; Sakal *et al.*, 1986; Jahiruddin and Cresser, 1991). Moreover Zn is known to form relatively stable chelates with organic ligands, which decreased their susceptibility to adsorption and fixation or precipitation. The incorporation of organic materials might have resulted in the formation of organic chelates of higher stability.

The correlation analysis also supported the above findings by registering negative correlation between the micronutrient availability and pH (r = -0.93* in the first year and r = -0.83** in the second year for DTPA- Zn, r = -0.93* in the first year and r = -0.93** in the second year for DTPA- Fe. The negative correlation of ESP (r = -0.96** in the first year and r = -0.93** in the second year) with DTPA- Zn and EAP (r = - 0.95** in the first year and r = -0.98** in the second year) with DTPA- Fe also lend support.

Rupa and Shukla (1998) reported that the availability of Cu in soil was mainly influenced by clay content, CEC and organic carbon. As discussed earlier, the added bio-resources improved the organic carbon content and CEC of the experimental soil and this might explain the reason for higher availability of Cu in the soil.

Many authors (Santhi *et al.*, 1998; Swarup and Yaduvanshi, 2000; Selvi *et al.*, 2003b) reported an increased availability of DTPA- extractable micronutrients in the rice soils due to the application of various bio-resources.

Table 4. Effect of bio-resource treatments on the DTPA- extractable micronutrients (mg kg^{-1}) of the soil in the field experiment-I

	Fe			Mn			Cu			Zn		
Treatments	AT	PI	Har	AT	PI	Har	AT	PI	Har	AT	PI	Har
T_1	7.41	6.08	6.45	7.04	6.52	6.13	0.71	0.65	0.61	0.69	0.64	0.61
T_2	11.40	10.52	9.28	10.11	9.50	8.18	0.92	0.84	0.75	0.95	0.85	0.77
T_3	10.36	9.27	8.80	9.64	8.83	7.38	0.88	0.79	0.73	0.90	0.79	0.72
T_4	9.92	8.84	8.18	9.27	8.10	7.33	0.81	0.74	0.70	0.86	0.74	0.67
T_5	7.83	7.60	7.58	7.92	7.16	7.07	0.75	0.69	0.66	0.70	0.67	0.65
SEd	0.34	0.40	0.38	0.17	0.33	0.34	0.03	0.02	0.02	0.02	0.01	0.02
CD (p=0.05%)	0.68	0.79	0.76	0.34	0.65	0.68	0.05	0.04	0.04	0.03	0.02	0.04

T_1- Control (100% NPK), T_2- 75% NP + 25% N through *Sesbania rostrata* + 25% P through Super Phosphate enriched FYM + K, T_3- 75% NP + 25% N through *Sesbania rostrata* + *Phosphobacterium* + K, T_4- 75% NP + *Azospirillum* + 25% P through Super Phosphate enriched FYM + K, T_5- 75% NP + *Azospirillum* + *Phosphobacterium* + K

Table 5. Effect of bio-resource treatments on the DTPA- extractable micronutrients (mg kg^{-1}) of the soil in the field experiment-II

	Fe			Mn			Cu			Zn		
Treatments	AT	PI	Har	AT	PI	Har	AT	PI	Har	AT	PI	Har
T_1	6.52	6.03	5.76	6.17	5.84	5.68	0.60	0.52	0.47	0.62	0.54	0.50
T_2	13.60	12.11	10.16	11.12	10.58	9.22	0.98	0.87	0.79	1.04	0.92	0.83
T_3	12.88	11.92	9.89	10.86	10.08	9.03	0.92	0.80	0.81	0.95	0.88	0.77
T_4	11.65	10.54	9.05	9.27	8.32	7.45	0.85	0.76	0.72	0.91	0.84	0.72
T_5	8.61	7.96	7.64	7.35	7.26	7.13	0.79	0.72	0.68	0.82	0.73	0.67
SEd	0.46	0.43	0.37	0.13	0.14	0.07	0.03	0.03	0.02	0.02	0.02	0.02
CD (p=0.05%)	0.92	0.86	0.73	0.25	0.27	0.14	0.05	0.04	0.03	0.04	0.03	0.04

7

Bioresources & Soil Biological Properties

The bio-resources application in the soil is known to improve the biological properties of the soil. Several earlier reports show a spur in microbial populations or increased microbial biomass carbon and the enhanced enzymatic activities with the incorporation of bio-resources in soil (Perucci, 1990; Goyal *et al.*, 1993; Bonde *et al.*, 2004).

Microbial populations

Indigenous microbial populations in soil are of fundamental importance for an eco system functioning, through determining nutrient cycling, organic matter decomposition and energy flow (Doran and Zeiss, 2000).

Subramani (1986) noticed that application of biofertilizers increased the soil microbial population such as bacteria (80%), fungi (60%) and actinomycetes (53%) upto 65 days after incorporation of organic manure. However, after 65 days a reverse trend was observed particularly 25 per cent decrease in bacteria, 24 per cent in fungi and 24 per cent in actinomycetes.

According to Nelidov *et al.* (1998), organic manures stimulate soil microorganisms thereby increases the productivity. Kukerja *et al.* (1991) reported that addition of organic manures increased microbial population under tropical conditions.

Goyal *et al.* (1993) observed an increase in the microbial biomass due to the addition of farmyard manure. The microbial biomass carbon (MBC) increased with the rate of application (upto 90 t ha^{-1}) and time (Increased upto 3 months and reverse trend observed thereafter).

A rapid increase in the microbial biomass carbon following the incorporation of cereal and legume residues were reported earlier by Goyal *et al.* (1994) and Azmal *et al.* (1996). Kumar and Kapoor (1995) showed that addition of gypsum and organic amendments reduced the soil pH and increased the microbial populations.

The addition of composts has a promoting effect on soil microbiology. Along with the content of organic matter of soil, the microbial populations usually increase. Chakrabarti *et al.* (1998) reported a rise in microbial biomass carbon in soils in consequence of compost fertilizer application.

Increase in soil microbes due to the increased availability of nitrogen with the application of organic residues and FYM was earlier reported by Bonde *et al.* (2004).

Research evidences also indicate increased microbial population under organic management of soil system than conventional management. Increased microbial populations and fertility status of soil under organic soil management was reported by Crecchio *et al.* (2004)and Pankhurst *et al.* (2005).

Through a laboratory incubation experiment, Sridevi *et al.* (2003) demonstrated that addition of prunings of *sorgham bicolor* or *Gliricidia sepium* @ 60 kg N equivalent to the soil increased the microbial biomass carbon and remained higher even after 90 days after incubation.

Rajendra Bhanwaria *et al.* (2022) reported that the microbial biomass build-up of C_{mic} increases by 43.13% over the control and 14.36% over the FYM. Similarly, the soil microbial biomass of $N_{mic,}$ and P_{mic} also increase considerably.

Soil enzymatic activities

Soil is an important natural resource that needs to be preserved and if possible, its quality and productive capacity improved. Enzyme activity in the soil is considered to be a major factor contributing to overall soil microbial activity and soil quality (Visser and Parkinson, 1992). The activities of soil enzyme are important in ensuring the availability of nutrients to plants. It is believed that most soil enzymes originate from soil bacteria, fungi and plant roots (Tarafdar *et al.*, 1988). Soil enzymes play a significant role in mediating biochemical transformations involving organic residue decomposition and nutrient cycling in soil (Martens *et al.*, 1992; Sarapatka and Krskova, 1997; McLatchey and Reddy, 1998).

Use of organic manures, green manures, crop residues along with inorganic fertilizers not only reduces the demand of inorganic fertilizers but also increases the efficiency of applied nutrients due to their favorable effect on physical, chemical and biological properties of soil (Pandey et al., 2007). María Alcívar et al.(2018) proved that soil microbial biomass increased 112% and 322% in the biochar+ humic substances + gypsum application.

Urease

Urea hydrolysis in soil is of great agricultural interest. The study of urease activity in soils constitutes a good model for the study of enzyme accumulation and stabilization in soils (O' Toole, 1991). It is also known that soil urease activity is conditioned by soil pH (Zantua and Bremner, 1977) and organic carbon (Bremner and Mulvancy, 1978) through the modification of native population dynamics (Kanivets *et al.*, 1971). Most of the earlier studies on soil urease had indicated that urease activity was highly correlated with organic carbon (Mulvaney and Bremner, 1981). Saline and alkali soils are usually low in available nitrogen and organic matter and coupled with high salinity and alkalinity, the urea hydrolysis and nitrification rates delayed due to poor bacterial activities in the soil. Subramani and Kannaiyan (1989) observed that urease activity was increased 73 per cent in bio fertilizer (Azolla) treated soil during first 65 days.

Perucci (1990) reported that the addition of municipal solid waste compost increased the activity of urease and reached maximum in three months. Subsequently, the activity showed a tendency to return to baseline values after reaching the maximum.

Kumar and Kapoor (1995) revealed that the amendment of alkali soils with gypsum or organic matter relieved pH stress and increased urease and dehydrogenase activities.

Sundaravarathan and kannaiyan (1998) found that the incorporation of *Azolla microphylla* or *Sesbania rostrata* along with the prilled urea in the soil resulted in the highest urease activity and total microbial biomass under submerged soil conditions. The soil native microbial population or the inocula introduced in the soil generally influences urease activity (Perotti and Pidello, 1999).

Masciandaro *et al.* (2004) revealed that the application of wheat straw @ 4 t ha^{-1} or dairy manure compost @ 35 to 70 t ha^{-1} to soil increased the urease activities considerably.

Phosphatase

Phosphorus that is associated with organic matter constitutes a significant proportion (80%) of total P present in the soil solution (Ron Vaz *et al.* 1993). Organic P has a biological origin and is derived primarily from microorganisms, plant degradation or plant root exudation (Anderson, 1980; Brooks *et al.,* 1984). Much of the organic matter first undergoes enzymatic hydrolysis (phytases and acid and alkaline phosphatases) to inorganic phosphate. Inorganic P liberated by this mechanism is than available for plant uptake or for sorption / exchange reactions with the soil matrix (Tarafdar and Claassen, 1988).

Thanikachalam *et al.* (1984) observed that phosphatase activity was increased to 75% in biofertilizers treated soil during first 70 days of cultivation. After that a reverse trend was observed and there was a reduction in phosphatase activity nearly 25 per cent.

Martens *et al.* (1992) indicated that addition of organic matter maintained high levels of phosphatase activity in soil during a long-term study. Goyal *et al.* (1993) found that the addition of farmyard manure in the field soils increased the phosphatase activity. The activity increased upto 90 days and a decline occurred after that.

More rapid turnover of microbial biomass, phosphorus and increased activity of phosphatase were observed in the soil receiving farmyard manure regularly (Oberson *et al.*, 1996). He further observed that the activity of phosphatase has been found to be higher in organic and biodynamically managed soils than under conventional management and control. Medina *et al.* (2004) reported that phosphatase activity was higher in rhizosphere of organic amended soils compared with non-amended soil.

Parham *et al.* (2002) stated that phosphatase activity was stimulated by microbial activity. Wan and Wong (2004) reported a higher phosphatase activity with phosphate solubilizing bacteria.

Cellulase

Cellulase are group of enzymes that catalyze the degradation of cellulose, a polysaccharide built of b-1,4 linked glucose units. Cellulose is the most abundant organic compound in the biosphere, comprising almost half of the biomass synthesized by photosynthetic fixation of CO_2 (Eriksson *et al.*, 1990). Cellulose in soils is derived mainly from plant residues incorporated into soil and a limited amount is derived from fungi and bacteria in soils (Richmond, 1991).

Bolton *et al.* (1985) observed that cellulase activity was increased with green manure incorporation. During first 50 days of incorporation 72 per cent increase was noticed and trend reversed after that and nearly 25 per cent reduction in cellulase activity was noticed.

Richmond (1991) observed an increase in cellulase activity with application of organic manures. Garcia-Garrido *et al.* (1992) reported an increased cellulase activity was due to biofertilizers application.

Dehydrogenase

Dehydrogenase is an intercellular enzyme linked to the respiratory electron transport system (Bolton *et al.*, 1985). The oxyreductase enzyme (dehydrogenase) has been proposed as a measure of overall microbial activity (Masciandero *et al.*, 2001), since it is an intercellular enzyme related to oxidative phosphorylation processes. Dehydrogenase activity plays an essential role in the initial stages of oxidation of soil organic matter.

Boltan *et al.* (1985) observed that dehydrogenase activity was increased to 75 per cent in green manure treated soils in the first 50 days after the incorporation.

Research evidences indicate that the incorporation of various organic manures increases the activity of dehydrogenase. Increase in the dehydrogenase activity with the addition of FYM (Goyal *et al.*, 1993; Prakash *et al.*, 2002), municipal solid waste (Perucci, 1992), vermi-compost (Marinari *et al.*, 2000) and green manure (Sriramachandrasekaran and Vaiyapuri, 2002) were already reported.

Goyal *et al.* (1998) reported a significant increase in the soil dehydrogenase activity and soil microbial biomass due to the incorporation of *Sesbania aculeata* in the soil in every alternate year. Increased dehydrogenase activity with organic manure addition was also reported by Parham *et al.* (2002).

Masciandaro *et al.* (2004) revealed that the application of wheat straw @ 4 t ha^{-1} or dairy manure compost @ 35 to 70 t ha^{-1} to soil increased the dehydrogenase activities considerably. Kaushik *et al.* (2005) stated that more than two times of dehydrogenase activity over control was found with the addition of FYM into soil.

Zhu *et al.* (2010) stated that OM and total nitrogen increased by 13% with rice straw incorporation and soil microbial biomass carbon, N and P was increased by 18-43%, as compared to NPK application alone. Straw incorporation resulted in improved soil physical conditions and crop productivity.

Katkar *et al.* (2012) described that the organic manures (FYM) along with recommended dose of fertilizers found to be viable option in increasing soil organic carbon, nutrient turn over, enhancing microbial biomass, thereby improvement in availability of nutrients in soil, maintenance of soil quality

Leo Daniel Amalraj *et al.* (2013) reported that among the different organic inputs, FYM recorded maximum dehydrogenase activity of 6.6 mg g^{-1} h^{-1}. Application of 100 percent RDF + FYM @ 15 t recorded significantly highe dehydrogenase activity of 360.9 mg TPF g^{-1} T^{-1}, alkaline phosphate activity of 23.49 mg gPNP g^{-1} h^{-1} compared control (286 9 mg TPF g^{-1} T^{-1} and 17.80 mg gPNP g^{-1} h^{-1} (Bhatt *et al.,* 2016).

Rajput *et al.* (2019) studied the effect of different organic amendments on rice and soil nutrients. The results of the experiment showed that rice + wheat compost significantly recorded higher level of soil organic carbon (20.29 g kg^{-1}), microbial biomass carbon (665 μg g^{-1}), microbial quotient, available nitrogen (140 mg kg^{-1}), and enzymatic activities (dehydrogenase - 5.1 μg TPF g^{-1} day^{-1}, alkaline phosphatise - 300 μg pNP g^{-1} dw^{-1} h^{-1} and urease - 210 μg NH_3 g^{-1} dry soil h^{-1}). The application of rice compost alone also increased the dehydrogenase and SOC of rice soils.

One of the foremost objective of utilizing bioresources is to ensure better proliferation of soil microbes and enhanced biological activity in terms of enzymatic acitivies of soil. In all the pot and field experiments, soil microbial populations, viz., Azospirillum, Phosphobacterium, bacteria, fungi & actinimycetes and soil enzymatic activites namely urease, phosphatase, cellulose and dehydrogenase were estimated. The results are discussed here under in line with the suitable literatures.

Pot and field experiments

Microbial populations

Sustainable agriculture is based, in part on the efficient management of soil microorganisms to improve soil quality (Kremer and Li, 2003). Microbial community helps in agriculturally important soil microbiological process like cycling of nutrients, build up of organic matter and improving the soil physical condition.

The improvement in the microbial populations due to the addition of various bio-resources was well evidenced in the present study (Table 1,3&4). All the microbes (*Azospirillum*, *Phosphobacterium*, bacteria, fungi and actinomycetes) increased in number upto certain period to reach the peak (panicle initiation stage) and then a decrease towards the end were observed though not lower than noticed in the initial stage.

The application of *Azospirillum* along with enriched FYM recorded the highest *Azospirillum* population of 1.42×10^{-6} / g soil in pot and 1.10 and 1.30×10^{-6} / g soil in the first and second year of field experiments respectively at the panicle initiation stage of rice cultivation. The population of *Phosphobacterium* was found highest with the application of Phosphobacterium along with green manure by registering the *Phosphobacterial* population of 8.67×10^{-6} / g soil in the pot experiment and 7.50 and 8.15×10^{-6} / g soil in the first and second year of field experiment respectively at the panicle initiation stage. It is self explanatory that the inoculated organisms maintained higher population and furthermore, the presence of bio-resources helped in proliferation in large numbers by the way of constant supply of C for their energy and N for nutritional requirements.

The population of total bacteria, fungi and actinomycetes in the soil were stimulated by the addition of various kinds of bio-resources. Among them, the addition of 75 % NP + *Sesbania rostrata* + enriched FYM outnumbered all other treatments by registering 46.33×10^{-6} / g soil of bacteria, 15.00×10^{-5} / g soil of fungi and 17.33×10^{-4} / g soil of actinomycetes in the pot experiment and 42.50 and 47.75 $\times 10^{-6}$ / g soil of bacteria, 21.50 and 24.00×10^{-5} / g soil of fungi and 19.75 and 21.25×10^{-4} / g soil of actinomycetes in the first and second year of rice cultivation respectively. The other bio-resource treatments also significantly improved the microbial strength in the soil over the control treatment, the application of 100 per cent RDF.

In salt affected soils, it was often reported that the development of microbiota are hindered due to the pH stress developed as the result of salinity. In the present study, the application of various bio-resources reduced the pH stress (Kumar and Kapoor, 1995) and provided conditions of favorable soil environment for microbe's development.

The increased microbial populations can be ascribed to the fresh input of bio-resources, in which the labial fractions of organic matter (which constitute the macro organic, light and humic fractions) provided the major source of energy for soil microorganisms (Gaunt *et al.*, 1995). In addition to direct C inputs, the bio-resources also improved soil moisture through reducing soil surface evaporation as reported by Martens (2001), thus mitigating the disruptive effect of soil drying on microbes.

As suggested by Cook and Allan (1992), the component N and P of the bio-resources, in addition to providing C substrate may also have the direct effect on the quantity and activity of microorganisms in the soil. In the present study, 75 per cent of N and P through fertilizers and remaining supplied through bio-resources have produced healthy crops, whose root exudates would increased microbial growth and activity indirectly (Lynch and Whipps, 1990).

The reduced population of bacteria (including *Azospirillum* and *Phosphobacterium*), fungi and actinomycetes towards the end of all the experiments may be ascribed to the exhaustion of available C substrate (Goyal *et al.*, 1993) and mineralization of the organic carbon (Perucci, 1992).

Several authors (Singaravel and Prasath, 2004; Crecchio *et al.*, 2004; Pankhurst *et al.*, 2005) were also reported an increased soil microbial population due to the application of various bio-resources.

Gogoi *et al.* (2018) revealed that enriched compost @ 5t ha^{-1} applied fields recorded maximum population of bacteria 9.11 log cfu g^{-1} soil and fungal population of 6.11 log cfu g^{-1} soil were exhibited in the enriched compost @ 5 t ha^{-1} applied plots.

Table 1. Effect of bio-resource treatments on the microbial assay of the soil in the pot experiment

Treatments	*Azospirillum* ($\times10^{-6}$ / g soil)			*Phosphobacterium* ($\times10^{-6}$ / g soil)			Bacteria ($\times10^{-6}$ / g soil)			Fungi ($\times10^{-5}$ / g soil)			Actinomycetes ($\times10^{-4}$ / g soil)		
	AT	PI	Har	AT	PI	Har	AT	PI	Har	AT	PI	Har	AT	PI	Har
T_1	0.17	0.23	0.20	3.36	4.75	4.12	21.33	25.67	23.67	7.33	8.67	7.67	9.67	12.00	10.33
T_2	0.56	0.64	0.60	4.70	5.51	4.87	34.00	43.67	39.33	11.67	13.33	13.67	12.67	15.67	13.33
T_3	0.84	1.10	0.95	4.67	5.45	4.93	23.67	29.33	26.00	9.00	10.33	9.33	11.67	13.33	12.33
T_4	0.45	0.52	0.48	4.95	5.49	5.33	32.47	40.00	36.00	11.33	12.67	12.00	13.67	15.00	14.67
T_5	0.24	0.32	0.29	5.33	7.67	6.50	23.00	28.67	27.33	8.00	9.67	9.33	11.67	13.00	12.33
T_6	0.81	0.95	0.85	5.67	7.23	6.87	38.67	46.33	43.00	13.33	15.00	14.67	14.67	17.33	16.00
T_7	0.62	0.69	0.65	5.93	8.67	7.33	35.67	44.67	39.67	12.67	14.33	13.67	13.67	16.67	15.33
T_8	1.10	1.42	1.21	4.93	6.50	5.67	36.33	45.33	38.00	12.00	13.00	13.33	13.33	16.00	14.67
T_9	0.95	1.21	1.10	5.67	7.79	6.95	27.67	31.67	29.33	10.00	11.67	12.00	12.00	14.00	12.67
SEd	0.03	0.04	0.03	0.18	0.21	0.17	1.09	1.47	1.29	0.32	0.43	0.56	0.46	0.44	0.37
CD (p=0.05)	0.06	0.08	0.05	0.36	0.41	0.32	2.18	2.94	2.57	0.63	0.86	1.12	0.92	0.87	0.74

T_1 - Control (100% NPK), T_2 - 75% N + 25% N through *Sesbania rostrata* +PK, T_3 - 75% N + *Azospirillum* + PK, T_4 - 75% P + 25% P through enriched Farm yard manure + NK, T_5 - 75% P + *Phosphobacterium* + NK, T_6 - 75% N + 25% N through *Sesbania rostrata* + 75% P + 25% P through enriched Farm yard manure + K, T_7 - 75% N + 25% N through *Sesbania rostrata* + 75% P + *Phosphobacterium* + K, T_8 - 75% N + *Azospirillum* + 75% P + 25% P through enriched Farm yard manure + K, T_9 - 75% N + *Azospirillum* + 75% P + *Phosphobacterium* + K

Table 2. Effect of bio-resource treatments on the enzyme activities of the soil in the pot experiment

Treatments	Urease (mg NH_4- N /g soil/ 24 h)			Phosphatase (mg P- nitrophenol/ g soil/ hr)			Dehydrogenase (mg TTF/ g soil/ 24 hr)			Cellulase (mg DNS/ g soil/hr)		
	AT	PI	Har	AT	PI	Har	AT	PI	Har	AT	PI	Har
T_1	22.40	25.20	23.57	13.99	14.93	13.67	117.10	127.37	122.33	12.93	14.30	13.53
T_2	49.47	56.93	54.80	21.67	23.20	18.37	162.10	180.97	175.00	40.87	49.60	44.37
T_3	39.23	54.13	52.10	17.33	18.33	15.27	129.93	144.00	138.43	15.40	15.97	15.63
T_4	33.60	44.80	42.90	25.67	28.63	23.30	156.80	172.70	168.10	21.23	26.33	23.83
T_5	26.00	31.60	29.83	22.67	25.73	21.67	125.70	137.17	131.40	14.83	15.87	15.10
T_6	56.93	69.13	67.10	32.33	33.63	28.40	175.53	202.17	193.10	47.83	58.43	52.93
T_7	51.33	64.40	62.33	28.99	30.03	25.93	168.77	183.20	182.80	42.17	51.30	45.63
T_8	42.87	48.60	46.23	25.67	27.23	23.57	161.77	176.23	170.53	22.37	25.37	24.20
T_9	39.20	44.80	43.07	23.67	24.13	21.03	133.80	143.23	136.93	15.47	16.60	15.97
SEd	2.28	1.50	1.38	0.81	1.24	0.86	5.59	5.01	4.09	1.70	1.50	1.73
CD(p=0.05)	4.55	2.99	2.75	1.62	1.71	1.50	11.18	10.01	8.17	3.40	2.99	3.46

Table 3. Effect of bio-resource treatments on the microbial population of the soil in the field experiment- I

Treatments	Azospirillum ($'10^{-6}$ / g soil)			Phosphobacterium ($'10^{-6}$ / g soil)			Bacteria ($'10^{-6}$ / g soil)			Fungi ($'10^{-5}$ / g soil)			Actinomycetes ($'10^{-4}$ / g soil)		
	AT	PI	Har	AT	PI	Har	AT	PI	Har	AT	PI	Har	AT	PI	Har
T_1	0.17	0.32	0.23	2.75	3.75	3.00	23.25	27.00	25.75	7.75	13.25	10.50	11.00	13.25	11.75
T_2	0.32	0.52	0.45	4.00	5.25	4.50	31.75	42.50	37.50	12.50	21.50	17.25	15.25	19.75	15.00
T_3	0.25	0.45	0.38	5.25	7,.50	6.75	29.00	38.75	34.00	10.75	18.50	15.00	14.00	17.25	14.00
T_4	0.69	1.10	0.95	3.50	4.75	4.00	26.25	32.00	29.75	9.25	17.00	13.75	12.75	15.50	13.25
T_5	0.52	0.95	0.84	4.75	6.00	5.50	25.50	29.25	26.50	8.50	15.75	10.25	12.00	14.25	12.75
SEd	0.03	0.04	0.03	0.21	0.25	0.32	0.98	1.60	1.32	0.46	1.25	0.91	0.61	0.62	0.49
CD (p=0.05%)	0.07	0.08	0.06	0.42	0.50	0.65	1.96	3.21	2.65	0.92	2.50	1.83	1.22	1.27	0.98

T_1- Control (100% NPK), T_2- 75% NP + 25% N through *Sesbania rostrata* + 25% P through Super Phosphate enriched FYM + K, T_3- 75% NP + 25% N through *Sesbania rostrata* + *Phosphobacterium* + K, T_4- 75% NP + *Azospirillum* + 25% P through Super Phosphate enriched FYM + K, T_5- 75% NP + *Azospirillum* + *Phosphobacterium* + K

Enzyme activities

Soil enzyme activity is a key feature of plant nutrition and cycling process, and therefore measurement of specific enzyme activities may be useful in determining the soil biological activity, which might be used as an indicator of soil fertility (Perucci, 1990). Most of the soil enzymes are believed that entirely derived from soil microbial population (Frankenberger and Dick, 1983; Angers *et al.*, 1993). Furthermore, enzymes catalyse many reactions in soil that are central to nutrient cycling, organic matter decomposition *etc.* (Dick, 1992).

The enzyme activities are reduced in salt affected soils. This indicates that biochemical quality is negatively affected in saline soils. This may be due to the 'salting-out' effect, which involves a decrease in enzyme solubility through dehydration, thus altering the enzyme catalytic site. The adverse effect depends on the concentrations of salts and the chemical composition of the enzyme itself (Garcia *et al.*, 2000). The enzymes get denatured under increased osmatic pressure of the soil solution.

In the present study, the enzyme activities (urease, phosphatase, dehydrogenase and cellulase) significantly and positively increased as the result of addition of various bio-resources (Table 2,5&6). All the enzyme activities estimated in the pot and field experiments showed increment upto panicle initiation stage of rice growth and the trend reversed towards the harvest.

Table 4. Effect of bio-resource treatments on the microbial population of the soil in the field experiment- II

Treatments	*Azospirillum* ($'10^{-6}$ / g soil)			*Phosphobacterium* ($'10^{-6}$ / g soil)			Bacteria ($'10^{-6}$ / g soil)			Fungi ($'10^{-5}$ / g soil)			Actinomycetes ($'10^{-4}$ / g soil)		
	AT	PI	Har	AT	PI	Har	AT	PI	Har	AT	PI	Har	AT	PI	Har
T_1	0.21	0.36	0.27	2.80	3.90	3.15	25.25	33.00	27.15	8.25	14.70	11.25	11.65	15.00	12.65
T_2	0.33	0.56	0.47	4.25	5.55	4.75	36.70	47.75	40.00	13.75	24.00	18.75	17.25	21.25	19.70
T_3	0.27	0.52	0.42	5.37	8.15	7.37	32.35	42.50	35.60	11.20	21.45	16.10	16.80	19.35	17.85
T_4	0.76	1.30	1.10	3.60	5.25	4.50	28.75	38.20	31.80	10.65	18.60	14.75	14.50	18.50	15.90
T_5	0.58	1.10	0.95	4.85	7.75	6.75	27.00	36.15	30.75	9.70	17.10	13.15	12.95	17.15	14.35
SEd	0.03	0.04	0.04	0.22	0.38	0.32	1.49	1.95	1.68	0.47	1.04	0.59	0.86	0.54	0.63
CD (p=0.05%)	0.05	0.07	0.08	0.43	0.75	0.64	2.97	3.89	3.36	0.93	2.08	1.17	1.72	1.08	1.25

Table 5. Effect of bio-resource treatments on the enzymatic activities of the soil in the field experiment- I

Treatments	Urease (mg NH_4- N /g soil/ 24 h)			Phosphatase (mg P- nitrophenol/ g soil/hr)			Dehydrogenase (mg TTF/ g soil/ 24 hr)			Cellulase (mg DNS/ g soil/hr)		
	AT	PI	Har	AT	PI	Har	AT	PI	Har	AT	PI	Har
T_1	26.50	37.25	30.15	12.00	18.50	15.25	112.70	135.00	124.10	12.60	17.75	13.85
T_2	42.75	58.75	49.50	24.70	32.75	27.60	152.00	187.50	167.75	28.15	45.50	32.65
T_3	38.00	52.50	42.65	21.25	24.50	23.45	144.20	169.75	152.00	22.10	41.75	25.00
T_4	32.15	47.50	36.00	18.00	21.25	20.40	135.10	158.00	140.80	19.45	28.50	22.10
T_5	29.55	42.00	31.75	15.15	21.75	16.70	127.50	147.50	132.00	15.15	21.00	18.40
SEd	1.88	2.30	2.83	0.99	0.88	1.04	5.83	6.89	5.13	0.96	1.10	1.30
CD(p=0.05)	3.75	4.60	5.65	1.97	1.75	2.08	11.65	13.78	10.26	1.92	2.20	2.60

Table 6. Effect of bio-resource treatments on the enzymatic activities of the soil in the field experiment- II

Treatments	Urease (mg NH_4- N /g soil/ 24 h)			Phosphatase (mg P- nitrophenol/ g soil/ hr)			Dehydrogenase (mg TTF/ g soil/ 24 hr)			Cellulase (mg DNS/ g soil/hr)		
	AT	PI	Har	AT	PI	Har	AT	PI	Har	AT	PI	Har
T_1	27.15	39.00	32.20	12.75	19.20	16.50	115.00	137.50	126.15	13.20	18.25	14.50
T_2	45.60	63.10	50.85	27.50	36.60	30.00	161.25	195.65	173.90	32.15	49.60	37.60
T_3	40.20	55.50	44.45	23.10	25.50	26.45	152.60	182.65	162.45	27.50	44.00	32.35
T_4	36.70	50.75	41.60	21.75	23.70	22.15	140.15	169.60	147.65	21.65	33.15	26.90
T_5	31.65	43.55	37.00	18.90	21.65	20.10	132.85	150.80	135.10	18.75	27.85	21.55
SEd	1.95	2.14	2.00	0.89	1.46	1.33	4.63	5.06	5.60	1.59	1.93	1.86
CD(p=0.05)	3.90	4.28	3.99	1.78	2.92	2.65	9.25	10.11	11.20	3.17	3.85	3.72

Urease

Alkali soils show low urease activity, but high pH does not affect the ammonifying or ureolytic bacteria as much as it affects the nitrifiers, which are highly sensitive to both salinity and alkalinity. Production of ammonia from urea may not be affected by salinity although nitrification may be severely affected by NaCl, Na_2SO_4, or $CaCl_2$. However, urease activity is reduced to some extent at higher pH but is enhanced with organic matter content of the soil (Rao and Pathak, 1996). In the present study also the results of pot and field experiment showed that the activity of urease increased with the application of various bio-resources. The application of 75% NP + green manure + enriched FYM recorded the highest urease activity of 69.13 mg NH_4- N /g soil/ 24 hr in pot experiment and 58.75 and 63.10 mg NH_4- N /g soil/ 24 hr in the first and second year of rice cultivation respectively.

The increased urease activity with the bio-resources application might be due to biomaterials added to the soil as well as the root exudates promoted the nitrogenase substances (Garcia *et al.*, 1994), which can induce the synthesis of urease in the salt affected soil. The results of the present findings are agreeable with the results obtained by Speir *et al.* (1980).

Sundaravarathan and Kanniyan (1998) and Masciandaro *et al.* (2004) were also reported an increased urease activity due to the application of different kinds of bio-resources.

Phosphatase

The alkaline phosphatase estimated in the pot and field experiments showed an increase with the addition of various bio-resources. Among the bio-resource treatments, the application of 75% NP + green manure + enriched FYM excelled in increasing the phosphatase activity of the soil. This treatment recorded 33.63 mg P-nitrophenol/ g soil/ hr in the pot experiment and 32.75 and 36.60 mg P- nitrophenol/ g soil/ hr of phosphatase activity in the first and second field experiments respectively. The other bio-resource treatments namely green manure + *Phosphobacterium*, *Azospirillum* + enriched FYM and *Azospirillum* + *Phosphobacterium* were also significant in increasing the phosphatase activity of the coastal saline soil as compared to control.

The demand for P by plants and soil microorganisms may have been responsible for the stimulation in the synthesis of phosphatase enzyme (Garcia *et al*, 1994). Moreover, the process related to the degradation of organic matter may be monitored through hydrolysis, such as phosphatase. According to Rao and Tarafdar (1992), an increase in the phosphatase activity indicates the changes and quality in soil phosphoryl substrates. The supply of readily metabolisable C in the organic byproducts is likely to have been the most influential factor contributing the increased phosphatase activity.

As reported by Nannipieri (1994), phosphatase activity increases when glucose and nitrogen sources are added to the soil, in other words, when the source of nutrients has an equilibrated balance between N and C (narrow C: N ratio) as in the applied green manure and enriched FYM.

The increase in the soil phosphatase activity with the addition of bio-resources could have been due to the soil substrate enrichment (priming effect) caused by the addition of mineral nitrogen to the soil. The phosphates added through bio-resources and fertilizer improved the phosphatase activity, which may be ascribed to the stabilized extra-cellular fraction of enzyme (Nannipieri, 1994). A significant positive correlation of phosphatase activity with organic carbon content (r = 0.92* and 0.88* in the first and second year respectively) and available P (r = 0.97** and 0.84** in the first and second year respectively) in the field experiments lend support to the present finding.

Many authors in the past (Oberson *et al.*, 1996; Parham *et al.*, 2002; Medina *et al.*, 2004) also reported the positive effect of applied bio-resources in increasing the phosphatase activity in the soil.

Dehydrogenase

The oxyreductase enzyme (dehydrogenase) has been proposed as a measure of overall microbial activity (Masciandero *et al.*, 2001), since it is an intercellular enzyme related to oxidative phosphorylation processes. In the present study, the added bio-resources significantly increased the dehydrogenase activity of the soil in both the pot and field experiments. The highest dehydrogenase activity of 202.17 mg TTF/ g soil/ 24 hr in the pot experiment, 187.50 mg TTF/ g soil/ 24 hr in the first field experiment and 195.65 mg TTF/ g soil/ 24 hr in the second field experiment at panicle initiation stage was recorded with the soil incorporation of *Sesbania rostrata* and enriched FYM as the sources for 25 % N and P respectively.

According to Rao and Pathak (1996), the incorporation of organic amendments to soil stimulates dehydrogenase activity because the added material may stimulate microbial activity in the soil. Pauscal *et al.* (1998) and Liang *et al.* (2005) reported the increased dehydrogenase activity with the incorporation of organics, owing to increase in microbial activity of the soil. These findings lend support to the results of the present investigation that the bio-resource application increased the soil microbial populations. Also a significant positive correlation of dehydrogenase activity with organic carbon (r = 0.97** and 0.99**), total bacterial population (r = 0.99** and 0.98**), total fungal population (r = 0.99** and 0.97**) and total actinomycetes population (r = 0.98** and 0.98**) in the first and second year of field experiments respectively supports to the present finding.

The slow and progressive decrease in the dehydrogenase activity after panicle initiation stage was because of the fact that this intercellular enzyme had weak activity since it was adsorbed on clay and humic substances when it was outside the microbial cells (Perez- Mateos *et al.*, 1988).

Boltan *et al.* (1985), Masiandaro *et al.* (2004) and Kaushik *et al.* (2005) also reported the increased dehydrogenase activity due to the application of various bio-resources in the soil.

Gogoi *et al.* (2018) reported that enriched compost @ 5t ha^{-1} application recorded the highest Dehydrogenase activity of 229.26μg TPF g^{-1} soil day^{-1}.

Cellulase

Cellulose, starch and sucrose are the major constituents of higher plants and among the most plentiful compounds in nature. The degradation of these compounds plays a significant role in the biological cycling of C in nature. The degradation process is mainly of microbial decomposition, which is restricted on the saline soil due to prevailing high pH and salinity. The saprophytic fungi, which can secrete cellulase and other exo-enzymes directly into the environment (Sensabaugh *et al.*, 1981; Saddler, 1986), constitute a major group of decomposer community.

In the present study, the cellulase activity estimated in pot and field experiments showed an increase with the addition of various bio-resources. Among the bio-resources, the application of 75% NP + *Sesbania rostrata* + enriched FYM excelled in registering the highest cellulase activity in the soil. In the pot experiment, it recorded 8.43 mg DNS/ g soil/hr and in the field experiments it registered 45.50 and 49.60 mg DNS/ g soil/hr of cellulase activity at the panicle initiation stages of first and second year respectively.

The increased cellulase activity with the addition of bio-resources was mainly due to the reasons that first, the increased fungal and bacterial numbers with the application of bio- resources (Kshattriya *et al.*, 1992) and secondly, due to the positive, and generally significant correlation between the moisture content of the substrate and the activity of cellulase, suggesting the favorable role of moisture in the synthesis of cellulase (Speir and Ross, 1981). The bio-resources applied no doubt had very much influential effect on the moisture related substrate parameters. Significant positive correlation of cellulase activity with bacteria (r = 0.98** and 0.99**), fungi (r = 0.97** and 0.94**) and actinomycetes (r = 0.946* and 0.930*), registered in the first and second year of field experiment respectively confirms the definite role of microorganisms in enhancing the cellulase activity of the soil.

The initial phase of decomposition was characterized by large amount of soluble constituents of the bio-resources, which are assimilable by the microbes (Linkins and Sinsabaugh, 1990). The alteration in the microbiota and substrate availability resulted in the decreased cellulase activity at the later phase.

Richmond (1991) and Garcia- Garrido *et al.* (1992) also reported an increased cellulase activity in the soil due to the application of bio-resources.

8

Bioresources and Rice Production

In the global scenario, the increasing energy crisis and consequent escalation of chemical fertilizer prices along with increasing risk associated with soil degradation have reviewed interest in bio-resources for attaining soil health and sustainable crop production. The use of organic manures, bio-fertilizers and chemical fertilizers will augment the efficiency of combined application of nutrients to maintain a high level of soil productivity and rice production (Thakuria *et al.,* 1991; Prasad *et al.,* 1995). Several experimental evidences showed that the combination of various bio-resources with chemical fertilizers improved the rice production in sustainable way (Roy and Jha, 1987; Subba Rao, 1995; Mahapatra *et al.,* 1997; Suresh *et al.*, 2001).

Growth and yield of rice

Green manures

Subba Rao *et al.* (1994) revealed that flooding and leaching of salts followed by the application of green manures @ 5.6 t ha^{-1} in coastal saline soils resulted in the highest rice grain yield of 24 q ha^{-1}. The highest grain and straw yield obtained was with the treatment, 100 per cent NPK along with application of *Sesbania rostrata* at 100 kg N equivalent rate (Budhar and Palaniappan, 1996).

Application of 80 kg ha^{-1} N through *Sesbania aculeata* a week before transplanting of rice or substituting basal dose of 50 per cent N through green manuring plus 50 per cent through prilled urea were comparable and produced significantly higher grain and straw yield over the application of chemical fertilizers alone (Mahapatra *et al.,* 1996).

Balasubramaniyan and Veerabadran (1997) showed that application of 50 percent recommended N through inorganic fertilizers and 25 per cent through *Sesbania cannabinus* + seed and soil inoculation with *Azospirillum* produced the highest grain yield of 5.87 t ha^{-1}.

Santhi *et al.* (1998) assessed the extent of fertilizer requirement under integrated plant nutrient system for rice- rice- pulse cropping sequence. The quantity of chemical fertilizer that could be adjusted to the levels and sources of organic manures was evaluated to be 38 kg N, 13 kg P_2O_5 and 33 kg K_2O ha^{-1} for fertilizer with green manure; 10-12 kg P_2O_5 ha^{-1} for fertilizer with *Phosphobacterium* and 40 kg N, 26 kg P_2O_5 and 33 kg K_2O ha^{-1} for fertilizer with green manure + *Phosphobacterium*.

The combined application of organic manures like *Sesbania aculeata* (6.25 t ha^{-1}) or FYM (12.5 t ha^{-1}) with *Azospirillum* @ 2 kg ha^{-1} and application of 150 kg N ha^{-1} recorded significantly shorter period for 50 per cent flowering, highest number of productive tillers m^{-2}, filled grains per panicle, panicle length and grain yield (5282 and 5218 kg ha^{-1}) in two consecutive rabi seasons (Shanmugam and Veeraputhran, 2000).

Omarhattab *et al.* (2000) reported that integrated nutrient management of 50 per cent N through green manure and 50 per cent inorganic nitrogen supply was found to be the best for getting higher grain yield of 3.58 t ha^{-1}.

A 10 year-long experiment at two locations (Khanpur and Varanasi, India) suggested that the application of 50 per cent RDF and 50 per cent NPK supplied through green manure (*Sesbania sesban*) resulted in increased grain yield @ 153 kg ha^{-1} Y^{-1} at Khanpur and 109 kg ha^{-1} Y^{-1} at Varanasi. Interestingly, the grain yield further increased with the application of 50 per cent RDF and remaining substituted through green manuring (191 and 124 kg ha^{-1} Y^{-1} at Khanpur and Varanasi respectively) (Yadav *et al.*, 2000). Duhan and Mahendra Singh (2002) revealed that green manuring with sunhemp along with 120 kg ha^{-1} of N increased the grain (4.99 t ha^{-1}) and straw yields (7.48 t ha^{-1}) over other green manures and RDF alone.

Selvi *et al.* (2003b) reported that *Sesbania aculeata*, grown with a seed rate of 60 kg ha^{-1} and ploughed in *situ* at 45 days after sowing + application of 50 per cent recommended doses of N and K was able to produce comparable economic yield (4847 and 8729 kg ha^{-1} of grain and straw) with 100 per cent recommended N and K alone (4904 and 8827 kg ha^{-1} of grain and straw).

Yadav (2004) reported that in *situ* green manuring with *Sesbania* at the seed rate of 15 kg ha^{-1} and application of 180 kg ha^{-1} of N significantly increased the grain yield of rice and recorded 5981 kg ha^{-1} as against 5538 kg ha^{-1} of non green manuring land. Ravindra Kumar *et al.* (2006) found that application of 15 t ha^{-1}green manuring or 5 q ha^{-1} FYM along with 100% NPK gave better yield, which was comparable with 75 % NPK + green manuring or FYM application.

Yaduvanshi *et al.* (2013) reported that the combination of green manure or 10 t FYM ha^{-1} with 50% recommended dose (60 kg N, 13 kg P and 21 kg K ha^{-1}) gave a rice yield similar to obtained from applying the 100% recommended dose of NPK (120-26-42 kg ha^{-1}). A net saving of 60 kg N and 13 kg P ha^{-1} was obtained through green manure or 10 t FYM ha^{-1}.

Nehr *et al.* (2017) reported that the application of 70% NPKS fertilizers + 4 t ha^{-1} dhaincha (Sesbania rostrata) (T1) produced the highest grain yield (5.90 t ha^{-1}), the second highest yield (5.85 t ha^{-1}) was obtained from 80% NPKS + 2 t ha^{-1} dhaincha (T2) treatment. The grain yield increased by 31.2 to 86.3% over control depending on the treatments.

Pratap Singh *et al.* (2018) concluded that the growth attributes such as plant height (96.8 cm), number of tiller m^2 (332), LAI (4.22) and dry matter accumulation (1285.7 gm^{-2}) were found highest with the treatment T5 (50% RDF + 50% N through FYM) but remained at par with all the treatment where either 25 or 50% N was substituted through organic sources . The differences in growth attributes were statistically alike between 100 % RDF and treatments supplicated either 25 or 50% N through organic sources.

Farmyard manure

Bhumbla (1972) and Singh *et al.* (1980) showed that combined application of FYM and gypsum was much more effective in increasing the yields and improving the properties of alkaline soils. Bandyopadhyay and Bandyopadhyay (1984) reported that application of FYM made a significant increase in grain and straw yield of rice in low productive coastal saline soil.

Aich *et al.* (1996) found that the application of gypsum @ 0.5 t ha^{-1} followed by leaching and FYM @ 10t ha^{-1} along with RDF produced the highest grain yield as compared to non-FYM application in a saline soil. Application of FYM @ 10 t ha^{-1} and BGA @ 13 kg algal crust ha^{-1} along with 100 per cent RDF produced highest grain yield of 4.22 t ha^{-1} and straw yield of 5.12 t ha^{-1} (Mishra and Sharma, 1997). Modak and Chavan (1999) reported an increase in the grain and straw yield of rice due to the application of FYM along with 100 per cent RDF.

An economy of 30 kg N, 15 kg P and 15 kg K were noticed with the addition of 10 t FYM ha^{-1} and 10 kg algal crust ha^{-1} (Dixit and Gupta, 2000). This treatment also produced 19.2 % more grain and straw yield of rice over application of recommended fertilizers alone. Babu *et al.* (2001) obtained the maximum growth characters and grain and straw yield of 5.52 and 9.60 t ha^{-1} respectively with the application of 12.5 t ha^{-1} of FYM along with 100 per cent RDF.

Satheesh and Balasubramaniyan (2003) reported that application of 50% N through FYM (10 t ha^{-1}) and remaining through neemcake (3 t ha^{-1}) registered the highest grain yield in two successive years. In the first year it was found 8 % and in second year 11.2% than that of conventional farming with chemical N fertilizers. Malarkodi and Singaram (2003) observed the highest grain (6129 kg ha^{-1}) and straw (9250 kg ha^{-1}) yield with the combined application of FYM @ 12 t ha^{-1} + recommended P through 1:1 ratio of jhabua rock phosphate and single super phosphate +*Phosphobacterium* @ 2 kg ha^{-1}.

Suman Mehta *et al.* (2004) found that with the application of 2.5 t FYM ha^{-1} + paddy straw @ 2.5 t ha^{-1} + 100 per cent RDF significantly produced taller plants, higher dry matter accumulation, more number of effective tillers and increased the grain (35.4 q ha^{-1}) and straw (57.0 q ha^{-1}) yield. In a long-term experiment, Saleque *et al.* (2004) found that application of 5 t ha^{-1} FYM along with 100 per cent RDF produced 19 % more yield than the soil application of chemical fertilizers alone.

Khanda *et al.* (2005) reported that on an average, integrated use of NPK fertilizers in conjugation with 10 t FYM ha^{-1} Y^{-1} increased the system productivity of all the cropping systems studied (rice- potato- sesame; rice- mustard- sesame; rice- wheat- sesame) by 8.7 per cent. Satyanarayana et al.(2006) showed that he highest grain yield of rice was obtained with the application of farmyard manure at 10 t ha"[1] and inorganic fertilizer at 120 : 60 : 45 kg N, P_2O_5 and K_2O ha"[1]. The increased grain yield was due mainly to increased nutrient uptake and number of tillers, filled grains per panicle and 1000-grain weight.

Gill et al.(2018) reported that the highest grain yield (34.9±0.54 q ha^{-1}) was obtained with combined application of FYM and 50 per cent of recommended nitrogen (RN) followed by GM+FYM+OP (33.7 q ha^{-1}) and GM+FYM (33.4±0.99 q ha^{-1}). Straw yield (63.1 q ha^{-1}) increased significantly ($P<0.01$) in treatment where FYM combined with 50 per cent of RN was applied. Anil kumar Saxena et al. (2020) reported that Combined application of 2.5 kg Zn + 5 t FYM/ha increased the grain and straw yield of rice and wheat crops by 39.3, 34.2, 29.5 and 15.0 per cent over control, respectively.

Enriched organic manures

Enriching organic manures with fertilizers and applying in comparatively small quantity produces significant effect on crop growth (Vimal and Talashilkar, 1983). Roy and Jha (1987) concluded that application of single super phosphate mixed with cow dung as basal dose resulted in significantly higher yield of rice.

Kuppuswamy *et al.* (1991) reported that enrichment of wet sludge with gypsum increased the rice yields by 0.95 and 0.71 t ha^{-1} in kharif and rabi seasons respectively, whereas enriched farmyard manure increased the rice yield by 0.67 and 0.50 t ha^{-1}.

Sharma and Mittra (1991) revealed that there was a marked improvement in the growth and yield of rice due to the application of enriched manures.

Kuppuswamy *et al.* (1992) observed that gypsum enriched biogas slurry in combination with recommended doses of NPK increased grain yield by 82.43 per cent over control with considerable savings of NPK fertilizers. Poongothai *et al.* (1995) revealed that application enriched FYM to rice crop enhanced the grain yield by 5 to 6 per cent over the control.

Malarkodi and Singaram (2003) suggested that the combined application of Jhabua rock phosphate and single super phosphate in 1:1 ratio along with FYM @ 12.5 t ha^{-1} and *Phosphobacteria* significantly increased the grain yield and nutrient uptake by rice. Singh *et al.* (2003) reported that application of single super phosphate incubated with farmyard manure at 1:1 ratios to rice crop significantly increased the plant height, panicle length, panicle weight, grain and straw yields and P uptake.

Baishya *et al.* (2015) reported that the organic sources, poultry manure (2.5 t ha^{-1}) was found to be most profitable in terms of productivity, profitability and sustaining soil fertility. The crop receiving 2.5 t poultry manure ha^{-1} along with 75 kg N +16.5 kg P + 31.3 kg K ha^{-1} improved yield attributes and yield (6.03 t ha^{-1}) as well as nutrient uptake and crop profitability over other treatments.

Singh *et al.*(2006) reported that the application of 25% recommended dose of N (RDN) through pressmud and the rest 75% NPK through inorganic fertilizers to both crops increased the grain yield of rice and wheat by 51.1 and 56.2% respectively over no NPK treatment. Kyi Moe *et al.* (2019) investigated the effects of integrated organic and inorganic fertilizers on the growth and yield of *indica* rice variety Manawthukha and *japonica* rice variety Genkitsukushi At the same N level, $CF_{50}PM_{50}$ application in both rice varieties resulted in higher SPAD values, plant height and tiller number than CF_{100}. $CF_{50}PM_{50}$ containing total N more than 4% supplied synchronized N for the demands of the rice plants, resulting in maximum dry matter, yield and yield components. $CF_{50}CM_{50}$ and $CF_{50}CP_{50}$ treatments containing total N less than 4% resulted in lower yields which were similar to CF_{100}. These results indicated that integrating organic and inorganic fertilizers enhanced growth parameters and yields of Manawthukha and Genkitsukushi, while reducing the dose of chemical fertilizer.

Sankavi *et al.* (2020) showed that the highest plant height (159.01cm), number of tillers (604 no/m^2), root length (25.18cm), dry matter production (12188 kg ha^{-1}), panicles/m^2 (479 no/m^2), filled grain per panicle (479 no/m^2), grain and straw yield (4535 kg ha^{-1} and 7259 kg ha^{-1}) were recorded with the application of 500 kg ha^{-1} poultry manure enriched with 100% zinc and iron (37.5 and 30 kg ha^{-1}) along with recommend dose of fertilizers over control.

Azospirillum

Azospirillum is known to have a close associative symbiosis with the higher plant system. Plant response to inoculation with *Azospirillum* in cereals and non cereals are often reported in terms of increased root length and volume, plant height, tiller numbers, early flowering, more test weight, increased grain and biomass yields, higher nitrogenase activities and N uptake (Boddey and Dobereiner, 1988; Wani, 1990).

Inoculation of *Azospirillum* in seed and soil increased the plant height and productive tillers hill^{-1}, grain and straw yields (Gopalaswamy *et al.*, 1989).

According to Ramakrishnan *et al.* (1992), the combined application of *Azospirillum* in seed, nursery and in main field could save 25 per cent of fertilizer nitrogen with considerable increase in grain yield of rice.

Hedge and Dwivedi (1994) reported that 48 out of 105 trails in kharif and 31 out of 55 trails in rabi season recorded that inoculation of *Azospirillum* to rice can save 25-50 per cent of recommended N fertilizer.

Inoculation of *Azospirillum* with 40 kg N ha^{-1} increased rice yield and was approximately equal to that obtained with 60 kg N ha^{-1} alone without inoculation (Subba Rao, 1995).

Syed and Sree Ramulu (1995) reported that application of 75 kg ha^{-1} of N + green manuring @ 6.25 t ha^{-1} + *Azospirillum* recorded the highest plant height, leaf area index, dry matter production, productive tillers, panicle length, number of filled grain and grain and straw yield of rice over the application of chemical fertilizers alone.

Application of *Azospirillum* along with 75 per cent N recorded increased grain yield of 5.87 t ha^{-1} than 100 per cent inorganic fertilizers which yielded 3.86 t ha^{-1} (Balasubramaniyan and Veerabadran, 1997).

Jayabal *et al.* (1999) observed that the combined application of FYM with 100 % recommended dose of fertilizers and *Azospirillum* inoculation significantly increased the yield and recorded the highest uptake of N, P and K by transplanted rice. Shanmugam and Veeraputhran (2000) reported that application of green manures and *Azospirillum* significantly increased the grain yield of rice.

Hemalatha *et al.* (2001) indicated that the application of 50 per cent of the inorganic N + *Azospirillum* recorded the highest grain yield of 6437 kg/ha^{-1} and straw yield of 8528 kg ha^{-1}.

Field experiment in a coastal saline soil with graded dose of N indicated that 50 per cent of recommended N dose (50 kg ha^{-1}) integrated with *Azospirillum* produced highest grain yield of 36.6 Q ha^{-1} (Kuldip Singh and Aulahb, 2001).

Vanaja and Sreenivasa Raju (2003) reported that in an integrated nutrient management studies for rice-sunflower cropping system, application of *Azospirillum* @ 2 kg ha^{-1} along with 75% of the recommended dose of fertilizers produced higher grain yield of rice and benefit cost ratio.

Phosphobacterium

Phosphobacteria are more efficient in the secretion of organic acids such as formic, acetic, propionic, glucolic, fumaric and succinic acids, which are responsible for dissolution of bound form of phosphates by altering the microenvironment of the soil (Gerresten, 1948). Addition of organic manures helps in increasing the solubilizing power of microorganisms.

Inoculation of rice seedling roots with phosphate solubilizing microorganisms increased the yield of paddy by 566 kg ha^{-1} as compared with non-inoculated seedlings (Gaur, 1990). Dhumsingh (1994) implied that without P application to the soil, *Phosphobacteria* increased the grain yield of rice by 10 to 20 per cent as compared with phosphorus application @ 60 kg ha^{-1} being only 5 to 15 per cent. Jayabal *et al.* (1999) reported that application of 100 % NPK fertilizers + FYM @ 12 t ha^{-1} + *Azospirillum* + *Phosphobacterium* significantly increased the yield and NPK uptake in rice.

Suresh *et al.* (2001) reported that application of 75 % P as enriched FYM along with *phosphobacteria* to seed and soil produced highest grain (2236 kg ha^{-1}) and straw (3678 kg ha^{-1}) yield of rain fed rice and saving of 25 % the chemical fertilizers. This treatment recorded the maximum growth (plant height) and yield (number of productive tillers, panicle length, number grains per panicle and 1000 grain weight) attributes of rice crop.

Kathirvelan and Vaiyapuri (2002) reported that 75 per cent N + green manure + *Azolla* and *Phosphobacterium* application increased the yield of rice by 21.17 per cent over the control.

Selvi *et al.* (2003a) reported that the application of *Phosphobacteria* @ 2 kg ha^{-1} along with green leaf manuring with *Pongamia glabra* and DAP increased the grain yield of rice, total P uptake and Phosphorus use efficiency.

Laxminarayana (2005) studied the effect of P solubilizing bacteria in soil and found that the application of 60 kg ha^{-1} rock phosphate in the soil along with inoculation of P solubilizing bacteria increased the P availability by 17.4 % over the application of rock phosphate alone.

Nuni Gofer et al.(2014) showed that the best treatment in term of plant height at 8 weeks after planting (WAP), the maximum number of tillers, number of productive tillers, number of grains per panicle and weight of milled dry rice were obtained in combination of 75% inorganic fertilizer and 300-400 kg ha^{-1} biofertilizer.

Neni Marlina *et al.* (2014) showed that the best treatment in term of plant height at 8 weeks after planting (WAP), the maximum number of tillers, number of productive tillers, number of grains per panicle and weight of milled dry rice were obtained in combination of 75% inorganic fertilizer and 300-400 kg ha^{-1} biofertilizer.

One pot experiment involving various bioresource treatments as presented in Table 1 and two field experiments involving treatments namely T_1- Control (100% NPK), T_2- 75% NP + 25% N through *Sesbania rostrata* + 25% P through Super Phosphate enriched FYM + K, T_3- 75% NP + 25% N through *Sesbania rostrata* + *Phosphobacterium* + K, T_4- 75% NP + *Azospirillum* + 25% P through Super Phosphate enriched FYM + K, T_5- 75% NP + *Azospirillum* + *Phosphobacterium* + K were studied in RBD with four replications. Various growth and yield characters were recorded at critical growth stages of rice. The results emanated from the experiments are discussed in line with suitable literatures.

Pot and field experiments

Growth of rice

In the present investigation, the application of all the bio-resources along with 75 per cent N and P and 100 per cent K through inorganic fertilizers had a pronounced effect on the growth characters of rice at all the stages of rice cultivation (Table 1-3). In promoting the overall growth of rice in the pot as well as the field experiments, the addition of 75 per cent NP + 25 per cent N substituted through *Sesbania rostrata* and 25 per cent P through enriched FYM outperformed all other treatments in producing tallest plants, highest dry matter production and number of tillers per plant.

With regard to plant height, in both the pot and field experiments the application of 75 per cent NP + *Sesbania rostrata* + enriched FYM excelled in increasing the plant height. The application of other bio-resources viz. green manure + *Phosphobacterium*, *Azospirillum* + enriched FYM and *Azospirillum* + *Phosphobacterium* also had significant effect in increasing the plant height over the control treatment, clearly indicating that a partial substitution of N and P had favorable effect on increasing the plant height. The results obtained was in accordance with the findings of Reddy *et al.* (1987) and Rajni Rani and Srivastava (2001), who demonstrated the positive influence of organic sources on the plant height.

Table 1. Effect of bio-resource treatments on the growth and yield of rice in the pot experiment

Treatments	Plant height (cm)			DMP (g / pot)			No.of tillers/ hill	No.of panic -les	Grain yield (g / pot)	Straw yield (g / pot)
	Till	PI	Har	Till	PI	Har				
T_1	38.99	60.59	62.70	26.58	31.63	35.65	5.66	4.99	55.82	84.36
T_2	44.25	67.15	70.77	32.19	39.46	44.06	8.44	6.55	65.25	91.35
T_3	42.11	63.32	67.82	30.38	37.83	40.69	7.44	5.77	61.17	85.64
T_4	43.12	63.41	68.06	30.30	37.80	40.93	7.89	5.99	63.28	87.96
T_5	41.38	63.30	67.73	30.44	38.08	41.94	7.33	5.77	59.76	83.65
T_6	49.33	71.56	76.11	35.83	43.41	48.71	10.89	8.33	80.46	119.08
T_7	46.51	70.71	74.47	33.89	41.37	47.11	8.89	7.33	75.68	107.45
T_8	45.21	68.88	73.72	33.12	41.13	46.36	8.99	6.77	71.50	102.25
T_9	45.21	69.52	73.51	33.89	41.37	47.11	8.66	6.77	68.62	94.70
SEd	1.91	1.09	1.57	0.55	0.76	0.60	0.50	0.27	3.15	5.21
CD (p=0.05)	3.82	2.18	3.14	1.10	1.52	1.19	0.99	0.54	6.32	10.41

T_1 – Control (100% NPK), T_2 - 75% N + 25% N through *Sesbania rostrata* +PK, T_3 - 75% N + *Azospirillum* + PK, T_4 - 75% P + 25% P through enriched Farm yard manure + NK, T_5 - 75% P + *Phosphobacterium* + NK, T_6 - 75% N + 25% N through *Sesbania rostrata* + 75% P + 25% P through enriched Farm yard manure + K, T_7 - 75% N + 25% N through *Sesbania rostrata* + 75% P + *Phosphobacterium* + K, T_8 - 75% N + *Azospirillum* + 75% P + 25 % P through enriched Farm yard manure + K, T_9 - 75% N + *Azospirillum* + 75% P + *Phosphobacterium* + K

Table 2. Effect of bio-resource treatments on the growth of rice in the field experiment-I

Treatments	Plant height (cm)			DMP (kg ha^{-1})			No. of tillers plant^{-1}
	AT	PI	Har	AT	PI	Har	
T_1	39.83	57.72	61.51	2924	4768	9456	6.75
T_2	53.26	74.43	83.06	5333	7311	12335	11.25
T_3	50.42	68.00	77.72	4862	6828	11938	11.00
T_4	48.21	66.29	73.25	4439	6437	11525	8.75
T_5	47.56	64.38	67.50	3583	5560	10516	7.50
SEd	0.89	1.60	1.93	162.86	180.10	198.72	1.27
CD **(p=0.05)**	**1.78**	**3.20**	**3.89**	**325.70**	**360.00**	**397.00**	**2.55**

T_1- Control (100% NPK), T_2- 75% NP + 25% N through *Sesbania rostrata* + 25% P through Super Phosphate enriched FYM + K, T_3- 75% NP + 25% N through *Sesbania rostrata* + *Phosphobacterium* + K, T_4- 75% NP + *Azospirillum* + 25% P through Super Phosphate enriched FYM + K, T_5- 75% NP + *Azospirillum* + *Phosphobacterium* + K

Table 3. Effect of bio-resource treatments on the growth of rice in the field experiment-II

Treatments	Plant height (cm)			DMP (kg ha^{-1})			No. of tillers plant^{-1}
	AT	PI	Har	AT	PI	Har	
T_1	39.53	57.24	62.06	2773	46.93	9248	6.50
T_2	54.55	75.54	84.36	5924	7930	13149	12.75
T_3	51.47	69.67	78.51	5266	7265	12572	12.25
T_4	48.69	67.00	74.42	4483	6578	11777	9.50
T_5	47.25	65.19	68.71	3763	5774	10951	7.75
SEd	0.92	1.39	1.79	136.64	148.81	210.88	1.09
CD(p=0.05)	**1.84**	**2.78**	**2.58**	**273.28**	**298.00**	**422.00**	**2.18**

The increased dry matter production at all the stages of rice growth was well pronounced in both the pot and field experiments due to the application of bio-resources as the partial substitution of N and P. The addition of *Sesbania rostrata* as the substitution for 25 per cent N and enriched FYM as the source for 25 per cent P was more efficient in increasing the dry matter production. This treatment registered 12,335 and 13,149 kg ha^{-1} in the first and second year of the field experiments respectively. The other bio-resources treatments also had favorable effect in increasing the dry matter production over the control treatment. The increased DMP with the application of various bio-resources were earlier reported by Shanmugam and Veeraputhran (2000); Yogananda and Reddy (2004) and Yung- Yu Shu (2005).

The number of tillers per plant also significantly increased with the application of various bio-resources treatments. Among the bio-resources treatments, the application of green manure + enriched FYM along with 75 per cent NP recorded the highest number of tillers. Earlier Balasubramaniyam and Palaniappan (1990) and Sharma and Das (1994) also indicated the increased tiller number in rice due to the application of various bio-resources.

The applied bio-resources along with inorganic fertilizers have significantly promoted the growth of rice in the coastal salt affected soil by their effect on increased and sustained nutrient supply throughout the crop growth period. An appreciable quantity of nitrogen and phosphorus (75 per cent) and full dose of potassium were applied in the form of inorganic fertilizers. The nutrients supplied by the fertilizers were sufficient to support the early growth of rice plants (Yung-Yu Shu, 2005). The nutrient added by the chemical fertilizers gets depleted after certain period (Tester, 1989), while the bio-resources continue to provide the nutrients for a longer period of time. The bio-resources are known to release the N at slower rates than chemical fertilizers and supplied N at the later stages of crop growth (Yoshida, 1981). The slow release of P from phospho- humic complex resulted from the decomposition of enriched FYM, besides solubilizing the insoluble P fraction of the soil enabled the supply of phosphorus synchronizing the crop demand.

The combined effect of slow release of N from the green manure and P from the enriched FYM by creating the conductive soil rhizosphere environment promoted better growth of rice in the coastal salt affected soil.

Further, the bio-resource are well known to improve the root growth of rice, which is associated with the improvement of biological activity in crop rhizosphere supported by amino acids and some physiologically active substances in the bio-resource materials (Zhou and Luo, 1997; Prasert, 1997) and improving the physical and chemical environment of the rhizosphere soil (Boparai *et al.*, 1992), which might have increased root length and volume thereby harnessing the uptake of water and nutrients from the soil and in turn the growth of rice promoted.

The biofertilizers are mainly the producers of plant growth hormones like indole acetic acid (IAA) and gibberellins as well as polyamines and amino acids (Thuler *et al.*, 2003) in addition to increased availability of N and P in the soil. This might explain the increased growth of rice with the inoculation of *Azospirillum* and *Phosphobacterium* along with other bio-resources.

Yield parameters and yield of rice

Yield parameters

In the present study, the yield parameters like number of panicles per plant, panicle length and thousand-grain weight in the pot and field experiments increased due to the application of various bio-resources (Table 1,4&5). Among the bio-resources treatments, the application of 75% NP + 25% N substituted through *Sesbania rostrata* + 25% P substituted through enriched FYM excelled in increasing the yield attributes of rice. In the field experiments, this treatment recorded the highest number of panicles (10.50 and 11.25 per plant in the first and second year respectively), longest panicles (23.56 and 23.78 cm in the first and second year respectively) and highest 1000- grain weight (20.42 and 23.78 g in the first and second year respectively). The other bio-resource applied treatments viz. 75% NP + green manure + *Phosphobacterium*, 75% NP + *Azospirillum* + enriched FYM & 75% NP + *Azospirillum* + *Phosphobacterium* were also significant in increasing the yield parameters of rice over the control treatment.

The betterment in yield components of rice due to the application of various bio-resources might be due to the release of nutrients during the process of decomposition and its effect on the increased vigor, photosynthate accumulation and better translocation of photosynthates to sink. Several other authors (Syed and Sreeramulu, 1995; Khanda and Dixit (1995); Suresh *et al*., 2001; Suman Mehta *et al*., 2004) also reported the increased yield parameters of rice due to the application of various bio-resources.

Table 4. Effect of bio-resource treatments on the yield of rice in the field experiment- I

Treatments	No. of panicles plant^{-1}	Panicle length (cm)	1000 grain weight (g)	Grain yield (kg ha^{-1})	Straw yield (kg ha^{-1})
T_1	6.25	20.74	19.72	3825	5207
T_2	10.50	23.56	20.42	5166	6868
T_3	8.25	22.40	20.26	4827	6661
T_4	7.50	22.06	20.08	4537	6302
T_5	6.75	21.80	19.96	4268	5926
SEd	0.57	0.07	0.49	170.90	105.88
CD **(p=0.05)**	**1.13**	**0.14**	**NS**	**341.80**	**211.76**

Table 5. Effect of bio-resource treatments on the yield of rice in the field experiment-II

Treatments	No. of panicles plant^{-1}	Panicle length (cm)	1000 grain weight (g)	Grain yield (kg ha^{-1})	Straw yield (kg ha^{-1})
T_1	6.00	20.26	19.60	3710	5077
T_2	11.25	23.78	20.51	5528	7280
T_3	9.00	22.52	20.30	5068	6961
T_4	7.75	22.16	20.16	4718	6491
T_5	6.50	21.92	20.04	4353	6030
SEd	0.61	0.08	0.44	163.55	101.27
CD**(p=0.05)**	**1.21**	**0.16**	**0.87**	**327.10**	**202.54**

Yield

The first and foremost aim of the present investigation is to increase the rice productivity of coastal salt affected soil by improving the soil fertility using bio-resource technology. The grain as well as the straw yield obtained from the pot and field experiments revealed the profound influence of all the applied bio-resources in increasing the rice productivity in the coastal salt affected soil.

In the pot experiment, the highest grain yield of 80.46 g pot^{-1} and straw yield of 119.08 g pot^{-1} was obtained with the treatment that received 75% NP + 25% N substituted through *Sesbania rostrata* + 25% P substituted through enriched FYM. This was 23.10 and 29.79 per cent increase over the application of 100% RDF. Next to this, the application of *Sesbania rostrata* + *Phosphobacterium* along with 75% NP (75.68 and 107.45 g pot^{-1} of grain and straw yield respectively) increased the grain and straw yield of rice. The application of *Azospirillum* + enriched FYM, and *Azospirillum* + *Phosphobacterium* also increased the yield of rice. The lowest yield of rice was recorded with absolute control.

The results of the field experiments revealed the validity of bio-resources in increasing the rice yield as established in the pot experiment. In the first year of rice production (November 2003- February 2004), the highest grain (5166 kg ha^{-1}) and straw (6868 kg ha^{-1}) yield was obtained with the application of 75% NP + 25% N substituted through *Sesbania rostrata* + 25% P substituted through enriched FYM respectively. This was 34.32 and 30.49 per cent higher grain and straw yield respectively over the application of chemical fertilizers alone. The treatment, biofertilizer (*Phosphobacterium*) in the presence of organic manure (*Sesbania rostrata*) followed this treatment by registering the grain yield of 4827 kg ha^{-1} and straw yield of 6661 kg ha^{-1}. The other biofertilizer (*Azospirillum*) in the presence of organics (enriched FYM and *Phosphobacterium*) was also efficient in increasing the yield of rice in the coastal salt affected soil as compared to application of chemical fertilizers alone.

In the second year of rice cultivation (November, 2004 - February, 2005) also the same trend continued as that of first year. However, the yield improvement in the second year was more due to bio-resources and ranged from 13.18 to 43.73 per cent as compared to 10.97 to 34.32 per cent in the first year. The highest grain and straw yield was registered with the application of 75% NP + 25% N substituted through *Sesbania rostrata* + 25% P substituted through enriched FYM (5528 and 7280 kg ha^{-1} of grain and straw yield respectively), which signifies a yield increase of 43.73 per cent in grain and 45.60 per cent in straw respectively over the application of 100 per cent RDF. The application of *Sesbania rostrata* + *Phosphobacterium*, *Azospirillum* + enriched FYM and *Azospirillum* + *Phosphobacterium* were also increased the rice yield significantly over the control treatment.

The marked increase in the yield of rice due to the application of green manuring and enriched FYM might have resulted from the adequate availability and transformation of organic nitrogen during the reproductive and grain filling stages coupled with increased rate of photosynthates from source to sink (Shanmugam and Veeraputhran, 2000). Further, the increased and steady availability of N and P from added bio-resources (Malarkodi and Singaram, 2003; Yogananda and Reddy, 2004) synchronized the crop demand and better utilization of N and P for grain production through metabolic activities. A number of authors (Santhi *et al.*, 1998; Omarhatab, 2000; Singh *et al.*, 2003; Ravindra Kumar *et al.*, 2006 in green manure and Dixit and Gupta, 2000; Satheesh and Balasubramaniyan, 2003; Saleque *et al.*, 2004; Khanda *et al.*, 2005 in FYM) reported the usefulness of organic manures in increasing the sustainable rice production.

The *Azospirillum* inoculation resulted in better yield of rice due to its effect on nitrogen fixation and production of extra cellular IAA and tryphophan, a precursor of IAA, which facilitates the growth and yield of rice as reported by Christiansen-Weniger (1998). The increased rice yield due to the application of *Azospirillum* have already been well documented by Jayabal *et al.* (1999), Hemalatha *et al.* (2001), Kuldip Singh and Aulahb (2001) and Vanaja and Sreenivasa Raju (2003).

The favorable effect of Phosphobacteria on rice growth may be explained by the increased P availability through solubilization of insoluble phosphates, decomposition of phosphate rich organic compounds and production of plant growth promoting substances (Gaur and Sunita, 1999). Further, the production of number of organic acids and its chelating action with metal cations resulted in the effective solubilization of phosphates (Laxminarayana, 2005). The increased yield due to the inoculation of phosphate solubilizing bacteria was earlier reported by Subba Rao (1999), Suresh *et al.* (2001), Selvi *et al.* (2003a) and Laxminarayana (2005).

9

Bioresources and Major Nutrient Uptake

Organic manures

Anand Swarup (1992) stated that application of green manures along with rice husk increased the concentrations of N, P, Zn, Mn and Fe and decreased the Na concentration of rice in a reclaimed sodic soil. Kolambe and Patel (1994) reported that application of organic amendments along with phosphorus increased the uptake of phosphorus at different growth stages of rice.

Talashilkar and Chavan (1997) reported that incorporation of 10 t of *Glyricidia* ha^{-1} and application of 100 per cent RDF recorded the highest total N uptake by rice over three successive years. Mishra and Sharma (1997) found that application of 100 per cent NPK + 10 t FYM ha^{-1} resulted in highest total uptake of NPK by rice. Mahapatra *et al.* (1997) revealed that application of 80 kg N ha^{-1} through green manure alone or through combined use of green manure and urea increased the total uptake of N by rice over application of fertilizers alone.

Modak and Chavan (1999) noticed an increased uptake of N, P, K, Zn and Fe by rice grain and straw due to the addition of 100 per cent NPK along with incorporation of 10 t FYM ha^{-1}. Application of *Sesbania* green manure or incorporation of mungbeen residue along with 60 kg ha^{-1} fertilizer nitrogen resulted in higher N uptake by rice similar to that obtained with 120 kg N ha^{-1} thus saving 50 per cent of fertilizer N (Sharma and Prasad, 1999).

Dixit and Gupta (2000) reported that the addition of 10 t ha^{-1} of FYM +10 kg ha^{-1} of BGA algal crust along with 100 per cent RDF recorded highest NPK content in rice grain. The NPK content in straw and total uptake followed the same trend. Yaduvanshi (2001) reported that the integrated use of green manure (*dhaincha*) or 10 t FYM ha^{-1} significantly increased the yield and nutrient uptake of rice.

Duhan and Mahendra Singh (2002) studied the influence of various green manures and found that application of 100 per cent RDN and green manuring with *sunhemp* recorded the highest yield and micronutrient uptake of Zn (329g ha^{-1}), Cu (62 g ha^{-1}), Mn (3545 g ha^{-1}) and Fe (4010 g ha^{-1}).

Over 10 years of continuous cropping, the N uptake of Kar rice was enhanced by 14.8 per cent with application of FYM @ 12.5 t ha^{-1} and 13.5 per cent with green leaf manure (GLM) and P uptake by 24.5 and 22.7 respectively over the application of chemical fertilizers alone. The K uptake was increased by 42.8 per cent with the incorporation of FYM (Baskar, 2003).

Satheesh and Balasubramaniyan (2003) reported that application of 50 per cent N through FYM and remaining as neemcake resulted in significant increase in the N, P and K uptake by rice grain and straw. Singh Parihar *et al.* (2003) reported that the highest N uptake was with the addition of 100 per cent RDN along with 5 t FYM ha^{-1}.

Addition of DAP as phosphorus source + green manuring (Pongamia glabra @ 6.25 t ha^{-1}) recorded the highest P uptake of 33.80 kg ha^{-1} (Selvi *et al.*, 2003a). A highest total NPK uptake of 310 kg ha^{-1} by rice was recorded with the application of 5 t FYM ha^{-1} in addition to 100 per cent RDF (Saleque *et al.*, 2004). Rathore *et al.* (2004) found that application of full dose of recommended fertilizers and FYM @ 10 t ha^{-1} increased the grain and straw uptake of N, P and K.

Powar *et al.* (2004) revealed that the uptake of NPK by rice was enhanced with the application of 50 percent recommended dose of fertilizers + 50 per cent through green leaf manure or 75 per cent RDF + 25 per cent N through paddy straw incorporation.

Bandyopadhyay *et al.* (2004) reported that application of 50 per cent N through chemical fertilizer + 50 per cent N through green leaf manure increased the N uptake of rice cultivated in a coastal saline soil. Ravindra Kumar *et al.* (2006) found that application of 15 t ha^{-1} green manuring or 5 q ha^{-1} FYM along with 100% NPK resulted in highest NPK uptake and was comparable with 75% NPK + green manuring or FYM application.

Singh *et al.* (2006) demonstrated that the rice-equivalent yield and total uptake of N, P and K by rice-wheat system were highest with 25% RDN through pressmud + 75% RDF through inorganic fertilizers, followed by 50% RDN through pressmud + 50% RDN through inorganic fertilizers.

The application of farmyard manure increased the uptake of N, P, and K by 20, 12, and 9%, respectively. There were significant interactions between organic manure and fertilizer treatment. Maximum uptake of N, P, and K was observed at a fertilizer level of 120 : 60 : 45 kg N : P_2O_5 : K_2O ha^{-1} in combination with application of farmyard manure(Sathiyanarayana *et al.*,2006) Sabina Ahmmmed *et al.* (2014) reported that the highest uptake of nitrogen (83.5 kg ha) and phosphorus (35.7 kg^{-1}) by rice were obtained with combined application of 120 kg N + 39.3 kg P + 100 kg K along with 10 t FYM ha^{-1} . The combined application of chemical fertilizers and organic manure increased organic carbon (OC), total N, available P, K and S contents in post-harvest soil. The overall results indicate that the integrated use of chemical fertilizer and organic manure can help increase grain yield of rice without deteriorating soil fertility (Nehr *et al.*, 2017).

Gill et al.(2018) showed that among different nitrogen management treatments, the maximum N, P and K uptake in grain and straw were observed under treatment T8 (FYM +50 per cent of RN) viz. integrated nitrogen treatment followed by T11 (GM+FYM+OP) and T9 (GM+FYM).

Sai Ram et al. (2020) reported that the uptake of nitrogen (N), phosphorous (P) and potassium (K)by rice was more (85.6, 28.1 and 128.7 kg/ha respectively) with 75% RDN + 25% N through vermicompost.

Biofertilizers

Mandal *et al.* (1993) observed that application of various bio-resources like FYM, water hyacinth and biofertilizers (*Azolla*) in combination with inorganic N enhanced the uptake of N, P and K by rice.

Surendra Rao and Sitaramyya (1997) observed that nitrogen uptake by rice under organic sources was in the order of *Azolla*> *Glyricidia*> poultry manure> FYM> biogas slurry. At harvest, 25 per cent substituted either by *Glyricidia* or by *Azolla* had significantly increased the N uptake over all other treatments.

Application of *Azospirillum* along with 75 per cent N recorded increased grain yield and total nitrogen uptake than the application of 100 per cent inorganic fertilizers (Balasubramaniyan and Veerabadran, 1997). Pany *et al.* (1998) reported that application of 27.5 kg P ha^{-1} in 1:1 ratio of P fertilizer and *Phosphobacterium* increased the P uptake of rice.

Deepa Thomas and James Martin (1999) reported that integrated use of green manure + biofertilizers along with chemical fertilizers increased the uptake of NPK by rice over thc control.

Jayabal *et al.* (1999) reported that the combined application of FYM and 100 % recommended dose of fertilizers along with *Azospirillum* inoculation significantly increased the yield & recorded the highest uptake of N, P & K by transplanted rice.

Lal *et al.* (2000) reported that the inoculation of *Azospirillum* helped in root development, hormone supply and as a consequence the uptake of mineral nutrients by rice crop. Soil application of *Azospirillum* @ 2 kg ha^{-1} in combination with FYM @ 12.5 t ha^{-1} influenced the higher uptake of nutrients and also nutrient availability in *rabi* rice. (Shanmugam and Veeraputhiran, 2000). Similar findings were also reported by Prasad *et al.* (1995) and Sreedevi and Thangamuthu (1991).

Hemalatha *et al.* (2001) indicated that the application of 50 per cent of the inorganic N + *Azospirillum* recorded the highest uptake of N by rice crop. Tiwari *et al.* (2001) revealed that in a partially reclaimed sodic soil, application of 10 t FYM ha^{-1} and blue green algae (BGA) with recommended NPK increased the yield and nutrient uptake of rice.

Ghosh et al.(2014) showed that the uptake of N, P, K, S, Ca and Mg by both grain and straw of rice were statistically significant due to use of integrated nutrient management. The highest nutrient uptake was recorded by the combined application of cowdung @ 5 t ha^{-1} along with recommended chemical fertilizers based on IPNS.

Neni Marlina et al. (2014) proved thar a significantly higher uptake of NPK fertilizer was obtained by combining NPK fertilizer at 75% of crop requirement and 300 kg ha^{-1} dose of biofertilizer (Biofertilizer was made by enriching straw compost with N2 interceptor bacteria, phosphate solvent bacteria and growth stimulator bacteria isolated from swamp lowland in South Sumatra).

Singh *et al.* (2016) concluded that the significant increase in yield and nutrient uptake with combined application of 3.6 g kg^{-1} of soil of rice husk biochar and PGPR with 75% RDF of NPK and Zn. The PGPR inoculation showed significantly higher N, P and K uptake in grain (0.58, 0.13 and 0.22 g pot^{-1}, respectively) and straw (0.47, 0.11 and 0.58 g pot^{-1}, respectively) over un-inoculated.

Sarkar *et al.* (2017) reported that the higher N (11.20 mg g^{-1}) and P (2.80 mg g^{-1}) concentration in grain was found in rice straw along with IPNS-based chemical fertilizer application. The total NPK uptake by rice plant was significantly higher in rice straw along with IPNS-based chemical fertilizer compared to farmers practice.

Vijayaprabhakar *et al.* (2020) studied the utilization of paddy straw as a source of nutrients for succeeding rice. The maximum NPK uptake was observed in case of rice straw + bio mineralizer and cow dung slurry (108.8, 39.1 and 128.2 kg ha^{-1}, respectively) and rice straw and cow dung slurry (104.6, 37.9 and 124.7 kg ha^{-1}, respectively) which were at par with each other.

Senthamizhkumaran, *et al.* (2021) reported that RDF 75% + Vermicompost @ 5 t ha^{-1} + Seed treatment with Azospirillum and Phosphobacteria + Soil application of AM fungi (biofertilizer enhanced the N availability and solubilized the inaccessible phosphorus, which thus recorded higher N accessibility and better phosphorus uptake when applied along with a recommended dose of fertilizer for rice.

Various bio-resources comprising *Sesbania rostrata*, *Azospirillum* for 25 per cent N substitution and enriched FYM and *Phosphobacterium* for 25 per cent

P substitution and their combinations along with control (100 per cent NPK) were studied under pot and field conditions. The results obtained with respect to major nutrient content and uptake by rice are discussed with available literatures.

Pot and Field experiments

Well established facts reveals that at a given level of salinity and sodicity, the growth and yield of crops are depressed more when the nutrition are disturbed than when it is normal (Bandyopadhyay and Rao, 2001; Kaushik *et al.*, 2005).

The nutrient availability and uptake by plants in salt affected soils depends on many factors in soils such as ionic composition and quantity of salts, presence and absence of toxic / complementary or antagonistic ions, type and quantity of clay, CEC, pH, EC etc. It is generally observed that N and P contents in plant decrease with increase in soil salinity (Bandyopadhyay and Rao, 2001). Generally, the soil salinity and sodicity decrease the root volume and thus the uptake of nutrients gets depressed. The sustainable approach of nutrient management in the salt affected soil with the balanced use of organic, inorganic and biological fertilizers will have the benefit of alleviating the ill effect of salinity and sodicity and increases the nutrient availability by creating a favorable soil physical, chemical and biological environment. Further, bio-resources by promoting root growth (Boparai *et al.*, 1992) facilitates higher nutrient uptake by the crop plants.

In the present study also application of various bio-resources viz. green manure, enriched FYM, *Azospirillum* and *Phosphobacterium* enhanced the contents of major, secondary and micronutrients and uptake of these nutrients by rice in the coastal saline sodic soil. In general, the content of all the nutrients were found higher at vegetative stage than at harvest stage of rice growth.

Nitrogen

In both the field experiments, the N content in rice plants at tillering, panicle initiation and in grain and straw was recorded higher in bio-resources treated plots than the control (Table 1,3&4). The increase in N content in all the bio-resources treated plots was more pronounced in the second year of rice cultivation than in the first year. The highest N content of 1.21 and 1.24 per cent in grain and 0.87 and 0.90 per cent in straw of first and second year respectively was recorded with the application of 75% NP along with *Sesbania rostrata* as the source for 25 per cent N + enriched FYM as the source for 25 per cent P. The other bio-resource treatments including the soil inoculation of *Azospirillum* along with 75 per cent N significantly increased the N content of rice in both the years.

Table 1. Effect of bio-resource treatments on the major nutrients content (%) of rice in the pot experiment

Treatments	N				P				K			
	AT	PI	Grain	Straw	AT	PI	Grain	Straw	AT	PI	Grain	Straw
T_1	1.53	1.12	0.86	0.63	0.25	0.18	0.15	0.12	1.49	1.37	0.53	1.05
T_2	1.64	1.25	1.09	0.76	0.29	0.25	0.20	0.19	1.63	1.46	0.62	1.26
T_3	1.63	1.24	1.05	0.73	0.27	0.23	0.17	0.15	1.55	1.36	0.57	1.19
T_4	1.61	1.19	1.07	0.75	0.28	0.19	0.20	0.18	1.58	1.38	0.58	1.19
T_5	1.59	1.24	0.97	0.71	0.27	0.22	0.18	0.16	1.53	1.39	0.55	1.17
T_6	1.75	1.31	1.18	0.84	0.34	0.27	0.26	0.25	1.70	1.49	0.71	1.31
T_7	1.72	1.26	1.14	0.82	0.32	0.25	0.25	0.24	1.67	1.49	0.68	1.28
T_8	1.69	1.24	1.12	0.82	0.29	0.22	0.25	0.24	1.63	1.46	0.63	1.27
T_9	1.65	1.22	1.10	0.80	0.28	0.23	0.23	0.21	1.62	1.42	0.63	1.25
SEd	0.02	0.02	0.01	0.02	0.03	0.02	0.01	0.02	0.01	0.03	0.02	0.02
CD (p=0. 05%)	0.04	0.03	0.02	0.04	0.05	0.04	0.02	0.04	0.02	0.05	0.04	0.03

T_1 - Control (100% NPK), T_2 - 75% N + 25% N through *Sesbania rostrata* +PK, T_3 - 75% N + *Azospirillum* + PK, T_4 - 75% P + 25% P through enriched Farm yard manure + NK, T_5 - 75% P + *Phosphobacterium* + NK, T_6 - 75% N + 25% N through *Sesbania rostrata* + 75% P + 25% P through enriched Farm yard manure + K, T_7 - 75% N + 25% N through *Sesbania rostrata* + 75% P + *Phosphobacterium* + K, T_8 - 75% N + *Azospirillum* + 75% P + 25% P through enriched Farm yard manure + K, T_9 - 75% N + *Azospirillum* + 75% P + *Phosphobacterium* + K

Table 2. Effect of bio-resource treatments on the uptake of major nutrients (mg pot^{-1}) by rice in the pot experiment

Treatments	N				P				K			
	AT	PI	Grain	Straw	AT	PI	Grain	Straw	AT	PI	Grain	Straw
T_1	243.21	355.13	483.46	544.80	37.21	58.05	82.85	101.25	239.22	432.28	296.22	885.17
T_2	309.02	491.63	716.52	710.00	59.55	100.75	130.65	191.85	301.00	573.60	404.96	1123.65
T_3	284.05	432.30	641.83	622.76	54.69	84.55	103.72	128.46	277.98	516.93	342.75	993.46
T_4	295.43	452.28	685.00	659.92	45.45	71.90	126.60	175.87	277.25	520.40	397.50	1064.18
T_5	296.79	460.70	595.26	857.18	56.32	85.00	107.55	150.26	280.05	528.05	328.65	995.50
T_6	367.26	568.65	949.75	995.66	68.08	117.23	217.43	296.85	349.35	641.10	563.10	1547.25
T_7	330.05	519.33	865.90	889.26	62.70	105.45	189.10	257.27	323.65	614.60	514.80	1358.93
T_8	326.21	508.60	797.66	838.13	57.96	93.60	178.92	245.36	312.99	598.50	464.55	1285.10
T_9	312.29	505.50	752.73	758.10	60.18	96.00	164.55	199.16	307.41	585.20	439.00	1174.33
SEd	11.75	16.60	36.74	38.27	2.49	4.26	7.93	10.81	12.23	27.77	13.28	31.17
CD (p=0.05%)	23.50	33.20	73.47	76.54	4.97	8.52	15.86	21.61	24.46	45.54	26.55	62.33

Table 3. Effect of bio-resource treatments on the content of major nutrients (%) at critical stages of rice in the field experiment-I

Treatments	N				P				K			
	AT	PI	Grain	Straw	AT	PI	Grain	Straw	AT	PI	Grain	Straw
T_1	1.62	1.29	0.88	0.68	0.29	0.24	0.17	0.13	1.54	1.35	0.55	1.07
T_2	1.87	1.48	1.21	0.87	0.36	0.31	0.29	0.27	1.71	1.48	0.72	1.35
T_3	1.85	1.44	1.19	0.85	0.34	0.28	0.28	0.22	1.64	1.42	0.68	1.30
T_4	1.79	1.39	1.08	0.79	0.32	0.26	0.25	0.19	1.60	1.39	0.63	1.26
T_5	1.76	1.35	1.00	0.74	0.30	0.26	0.21	0.15	1.57	1.38	0.59	1.18
SEd	0.02	0.02	0.03	0.02	0.01	0.02	0.02	0.02	0.03	0.02	0.02	0.02
CD (p=0.05%)	0.04	0.04	0.05	0.04	0.02	0.04	0.04	0.05	0.05	0.03	0.04	0.04

T_1- Control (100% NPK), T_2- 75 % NP + 25 % N through *Sesbania rostrata* + 25 % P through Super Phosphate enriched FYM + K, T_3- 75 % NP + 25 % N through *Sesbania rostrata* + *Phosphobacterium* + K, T_4- 75 % NP + *Azospirillum* + 25 % P through Super Phosphate enriched FYM + K, T_5- 75 % NP + *Azospirillum* + *Phosphobacterium* + K

The N uptake at all the critical stages of rice growth was also increased with the application of various bio-resources as evidenced in both the pot and field experiments (Table 2,5&6). The highest N uptake of 949.75and 995.66 mg pot^{-1} in grain and straw respectively was registered with the application of 75 % NP + green manure + enriched FYM. This treatment also recorded the highest N uptake in the field experiments by registering the N uptake of 62.51 kg ha^{-1} and 59.77 kg ha^{-1} in the grain and straw of the first crop and 67.97 and 65.12 kg ha^{-1} in the grain and straw of the second crop respectively. The other bio-resource treatments viz. green manure + *Phosphobacterium*, *Azospirillum* + enriched FYM and *Azospirillum* + *Phosphobacterium* were also significantly increased the N uptake by rice over the control treatment.

The availability of nitrogen to crops is reduced in saline soils because of slow mineralization rate and dominant chloride anions (Hu and Schmidhalter, 1997). Bio-resources will have the benefit of increasing the mineralization rate by increasing the microbial activity and alleviating the toxic effect of chlorides. Bio-resources also serve as slow release source of nitrogen and also temporarily bind the ammoniacal pool resulting from the urea hydrolysis. Organic manures have the additional benefits on improving the soil physical properties along with reduction in loss of nutrients due to temporary bio fixation of nutrients by soil microbes or by clay humus complex as suggested by Bandyopadhyay and Bandyopadhyay (1984), besides root proliferation, resulting in a larger root surface available for the uptake of nutrients (Mandal *et al.*, 2003).

Fixation of atmospheric nitrogen is the original proposed mechanism by which *Azospirillum* increases the soil N concentration in cropped lands (Yoav Bashan, 2004). The rice plant also stimulates the nitrogenase activity of endophytic *Azospirilla* by providing carbon source as energy (Christiansen- Weniger, 1998; Burdman et al., 2001), inducing more N_2 fixation resulting in more N uptake.

Talashilkar and Chavan (1997), Modak and Chavan (1999), Dixit and Gupta (2000), Baskar, (2003), Singh Parihar *et al.* (2003), Powar *et al.* (2004) and Ravindra Kumar *et al.* (2006) also reported earlier the increased N uptake by rice due to the application of various bio-resources.

Phosphorus

A significant increase in the content and uptake of P by rice due to the bio-resources application was well evidenced in the present investigation. The P content in rice increased significantly in the range of 0.21 to 0.29 per cent in the grain and 0.15 to 0.27 per cent in straw of first year and 0.23 to 0.29 per cent in grain and 0.17 to 0.27 per cent in straw of second year field experiments (Table 3&4).

In tune with the P content, at all the critical stages, the uptake of P by rice was significantly increased by the bio-resources treatments. In all the experiments, the highest uptake of P by rice was recorded with the application of 75% NP + 25% N substituted through green manure and 25 % P substituted through enriched FYM (Table 2,5&6). In the pot experiment, it recorded 217.43 and 296.85 g pot^{-1} in grain and straw respectively. In the field experiments this treatment registered the P uptake of 15.07 kg ha^{-1} in grain and 18.65 kg ha^{-1} in straw in the first year and 16.13 kg ha^{-1} in grain and 19.69 kg ha^{-1} in straw of the second year respectively.

In the present study, the enhanced uptake of P by rice with the application of various bio-resources is due to the increased P availability in the soil by the decomposition of applied bio-resources (Yadvinder Singh *et al.*, 1988) as well as the synergistic effect existed between nitrogen and phosphorus. The adsorption and dissolution of P in the soil is greatly influenced by soil pH and organic acids (citric, acetic, tartaric, malic etc.) present in soils. The organic acids derived from plant roots and increased microbes might have helped in release of relatively non-labial P in the soils as reported by Pratap Bhattacharyya and Datta (2005). The release of organic acids and reduction in soil pH due to the bio-resource incorporation no doubt have the beneficial influence of increased uptake of P by rice. A significant negative correlation of P uptake with pH (r = -0.99** in the first year and r = -0.92* in the second year) and positive correlation with available P (r = 0.99** in the first year and r = 0.98** in the second year) and phosphatase activity (r = 0.99** in the first year and r = 0.98** in the second year) support the present findings.

The increased P availability with the inoculation of *Phosphobacterium* might be due to the ability to produce organic acids and solubilizing the insoluble phosphates in high pH soil (Laxminarayana, 2005). The influence of *Azospirillum* to olubilized the insoluble inorganic phosphate in the soil is also reported. Sugars, like glucose, that are part of root exudates enhance the capacity of *Azospirillum* to olubilized the insoluble $Ca_3(PO_4)_2$. Further, the acidification of rhizosphere by producing different organic acids by bacteria also increased the solubility of P containing substances (Seshadri *et al.*, 2000; Amooaghaie *et al.*, 2002). The duel inoculation of *Phosphobacterium* and *Azospirillum* brought synergistic effect (Yogananda and Reddy, 2004) and thus increased the P availability as a consequence the uptake of P by rice.

Yaduvanshi (2001), Satheesh and Balasubramaniyan (2003) and Ravindra Kumar *et al.* (2006) also reported earlier the increased uptake of P by rice with the addition of various bio-resources.

Table 4. Effect of bio-resource treatments on the content of major nutrients (%) at critical stages of rice in the field experiment- II.

Treatments	N				P				K			
	AT	PI	Grain	Straw	AT	PI	Grain	Straw	AT	PI	Grain	Straw
T_1	1.57	1.27	0.87	0.66	0.29	0.23	0.17	0.12	1.53	1.32	0.54	1.05
T_2	1.92	1.52	1.24	0.90	0.39	0.35	0.29	0.27	1.75	0.53	0.76	1.40
T_3	1.86	1.46	1.23	0.88	0.36	0.31	0.28	0.25	1.67	1.47	0.72	1.34
T_4	1.81	1.41	1.10	0.81	0.33	0.28	0.27	0.21	1.62	1.42	0.65	1.29
T_5	1.77	1.36	1.01	0.74	0.31	0.27	0.23	0.17	1.58	1.40	0.60	1.22
SEd	0.03	0.02	0.02	0.02	0.01	0.02	0.02	0.02	0.03	0.03	0.02	0.03
CD (p=0. 05%)	0.05	0.04	0.03	0.03	0.02	0.03	0.03	0.04	0.05	0.05	0.03	0.05

Table 5. Effect of bio-resource treatments on the uptake of major nutrients (kg ha^{-1}) at critical stages of rice in the field experiment- I

Treatments	N				P				K			
	AT	PI	Grain	Straw	AT	PI	Grain	Straw	AT	PI	Grain	Straw
T_1	47.58	61.53	34.00	35.41	8.42	11.34	6.47	6.27	44.86	64.28	21.05	55.74
T_2	99.67	108.27	62.51	59.77	19.22	22.56	15.07	18.65	91.05	107.92	37.23	92.70
T_3	89.46	98.41	56.49	55.25	16.55	19.08	13.46	14.59	79.14	96.79	32.80	87.30
T_4	79.43	89.27	48.89	49.80	14.24	16.67	11.72	11.28	71.08	89.43	28.57	79.42
T_5	62.98	75.00	42.26	43.81	10.71	14.53	8.98	8.90	56.20	76.28	25.16	69.88
SEd	3.06	3.84	2.51	2.49	0.68	0.71	0.77	0.85	2.96	3.53	1.13	1.60
CD (p=0.05%)	6.12	7.68	5.04	5.00	1.36	1.41	1.55	1.70	5.92	7.05	2.27	3.22

Table 6. Effect of bio-resource treatments on the uptake of major nutrients (kg ha^{-1}) at critical stages of rice in field experiment- II

Treatments	N				P				K			
	AT	PI	Grain	Straw	AT	PI	Grain	Straw	AT	PI	Grain	Straw
T_1	43.08	59.64	31.72	34.17	8.32	10.76	5.96	6.60	42.98	61.74	19.30	52.93
T_2	113.57	120.84	67.97	65.12	23.15	27.96	16.13	19.69	103.51	120.72	42.65	100.44
T_3	97.71	106.58	59.93	59.46	18.11	22.46	14.22	17.42	87.38	106.23	36.64	91.13
T_4	81.53	92.18	51.48	52.75	14.98	18.18	12.68	12.92	72.70	84.90	30.76	82.54
T_5	66.89	78.15	44.05	44.06	11.12	15.26	10.19	10.25	59.26	80.48	25.27	74.18
SEd	4.44	5.09	2.41	2.48	1.68	1.29	0.59	0.93	4.54	6.33	1.22	3.67
CD (p=0.05%)	8.89	10.17	4.82	4.95	3.36	2.57	1.18	1.86	9.08	12.65	2.43	7.33

Potassium

The content and uptake of K by rice was significantly and positively influenced by the application of various bio-resources. In the field experiments, the highest K content of 0.72 per cent in grain and 1.35 per cent in straw in the first year and 0.76 per cent in grain and 1.40 per cent in straw in the second year was recorded with the application of 75% NP + *Sesbania rostrata* + enriched FYM. The other bio-resources treatments were also found useful in increasing the K content in rice over no bio-resource application.

The results of pot as well as the field experiments indicated the increased uptake of K by rice with the application of various kinds of bio-resources (Table 2,5 &6). Among the bio-resource treatments, the soil incorporation of *Sesbania rostrata* @ 25% N equivalent and enriched FYM @ 25% P equivalent along with 75% NP resulted in higher K uptake by rice. This treatment recorded the highest K uptake of 563.18 mg pot^{-1} by grain and 1547.25 mg pot^{-1} by straw in the pot experiment. In the field experiments, this treatment registered 37.23 kg ha^{-1} by grain and 87.30 kg ha^{-1} by straw (first crop) and 42.65 kg ha^{-1} by grain and 100.44 kg ha^{-1} by straw (second crop) respectively.

Potassium in the organic materials is not strongly bound to any particular constituent and is readily available to the first and immediate crop to which it is applied (Srinivasa Rao *et al.*, 1996) through the process of mineralization. The production of organic acids with the incidence of decomposition of bio- resources helps in releasing K from exchangeable and non- exchangeable sites of clay and other soil fraction on prolonged submergence and the organic acids release (Mohanty and Patnaik, 1997). The increased K content and uptake in rice throughout its growth period might be due to the above-mentioned factors. The minimal loss of K through leaching as the result of fixation of K ions on the cation exchange sites might further enhanced the availability (Selvi *et al.*, 2003b) and in turn the greater uptake by rice.

The earlier records of Powar *et al.* (2004), Rathore *et al.* (2004), Saleque *et al.* (2004) and Ravindra Kumar *et al.* (2006) also implied the higher K uptake by rice due to the addition of various bio-resources.

10

Bioresources on Secondary and Micro Nutrient Uptake

The uptake of Mg by rice plant was highest when fertilizer was applied in combination with vermicompost (Jadhav *et al.*, 1997). Pattanayak *et al.* (2001) also reported the highest Ca uptake in the treatments that received green manuring of dhaincha. Sriramachandrasekharan (2001) who reported the highest Fe uptake of rice with green manuring of Sesbania aculeata.

Debele *et al.* (2001) reported that the application of FYM enriched with 10% rock phosphate + gypsum with PSB and PSF @ 500 g inoculums t^{-1} of FYM increased the nutrient uptake of maize. The higher nutrient uptake of 227 kg N ha^{-1}, 59.5 kg P ha^{-1} and 227 kg K ha^{-1} recorded with the application of enriched FYM @ 2 t ha^{-1}.

Organic sources of zinc were found to be significantly superior in increasing the zinc uptake compared to the inorganic sources. Microbial biomass and soil enzymes increases with application of organic amendments. The organic material can improve the availability of Zn by releasing Zn with time and through changes in physico-chemical properties of soil (Liang *et al.*, 2003).

Chander *et al.* (2007) who observed that the incorporation of FYM resulted in to a significant and consistent increase in Ca uptake by cauliflower.

Soil organic matter may increase the soluble /available fractions of Zn in soil for plant uptake. Moreover application of organic amendments also improves the biological properties of soil (Tajeda *et al.,* 2006). Mohana Rao Puli *et al.* (2017) reported that application of FYM 10t ha^{-1} + RDF had recorded 50.0-77.13% higher Ca uptake & 34.7-27.9% higher Mg uptake by rice grain over NPK control.

Application of $ZnSO_4$ and $FeSO_4$ fortified FYM significantly increases the availability of micronutrients. Application of Zinc through organic sources was found to increase the uptake of S, Cu,Mn,Zn and Fe significantly. An increase in uptake of above nutrients by maize due to addition of organic sources such as FYM, pressmud compost and vermicompost was probably because of supply of these nutrients through the organic sources (Kumar *et al.*, 2010).

Singhal *et al.* (2012) reported the higher uptake of micronutrients *i.e.* Zn, Fe, Mn and Cu by maize grain with the application of enriched compost with reduced dose of nitrogen. The highest uptake of micronutrients due to application of enriched compost @ 2 t ha^{-1} with 75% reduced dose of nitrogen might be attributed to the production of various organic acids of low molecular weight such as succinic, citric, malic, oxalic and acetic acids in addition to the release of N and micronutrient cations.

Nehr *et al.* (2017) reported that the application of 70% NPKS fertilizers + 4 t ha^{-1} dhaincha (*Sesbania rostrata*) (T_1) resulted in higher N, P, K and S uptake by rice. The N, P, K and S concentrations of rice grain was higher for the treatments where Sesbania and cow dung were together applied in combination with NPKS fertilizers.

Shormy *et al.* (2013) reported that the nutrient contents (N, P, K, S, Ca and Mg) and their uptake by grain and straw were significantly influenced by the application of different treatments. The highest values of most of the parameters were obtained from RD-NPKS + PSB treatment which was statistically identical with GM 5 t/ha +FYM 5 t/ha + PSB, GM10 t/ha and GM 5 t/ha + FYM 5 t/ha treatment combinations in many cases.

Uma Shankar Ram *et al.* (2013) Fertilizer source as application of 75% RFD through inorganic and rest through FYM recoded significantly higher growth, yield and N, P, K, Zn and Fe content of grain than 100% RFD through inorganic source.

Michalak *et al.* (2017) reported that the addition of seaweed compost increased the micro and macronutrient content such as B by 108%, Mn by 25% and 75% and Zn by 17%.

Patra *et al.* (2018) reported that the Zn, Fe, Cu and Mn uptake of grain and straw were maximum in 25% NP + 100% K + enriched compost @ 2 t ha^{-1}. The grain and straw nutrient content Zn - 120%, Fe - 181%, Cu - 143%, Mn - 138% and N - 128%, P - 386%, Zn - 196%, Fe - 14.8%, Cu - 190%, Mn - 110%, respectively was increased over control.

Zeb *et al.* (2018) reported that application of zinc solubilizing bacteria enriched compost @ 60:40 resulted in higher Zn release in soils when compared to $ZnSO_4$ application. The zinc accumulation was greater in shoot, root, and grain by 18, 16 and 11% compared to the application of zinc sulphate.

Anil kumar Saxena *et al.* (2020) reported that application of treatments combined application of 2.5 kg Zn + 5 t FYM/ha after two crop cycles resulted higher Zn uptake in rice grain as compared to treatment application limited to first year rice crop. Combined application of Zn and FYM was more effective in increasing total Zn uptake of rice and rice + wheat than individual application of FYM or

Zn. The interaction effect of treatments and their application schedule significantly influenced uptake of Cu, Fe and Mn in crops.

Santosh Yadav *et al.* (2022) indicated that the application of Zn enriched FYM increased the organic carbon and DTPA-Zn and Fe significantly.

Various bioresource treatments which were evaluated for the major nutrients uptake were also studied for the seconday and micro nutrients. The experimental results obtained with respect to secondary and micronutrients are discussed in line with suitable literatures.

Secondary nutrients

Sulphur

The increased S content and uptake by rice was realized in the present investigation due to the application of various bio-resources. The increased S content of rice in the field experiment due to the application of various bio-resources ranged from 0.13 to 0.20 per cent in grain and 0.18 to 0.28 per cent in straw of the first year and 0.15 to 0.23 per cent in grain and 0.19 to 0.30 per cent in straw of the second year respectively (Table 3&4). Among the treatments, the highest S content at all the stages of rice growth was recorded with the application of green manure and enriched FYM as the sources for 25 per cent N and P respectively.

The application of *Sesbania rostrata* and enriched FYM along with 75% NP proved its superiority over all other treatments in enhancing the S uptake by rice and was exhibited in all the experiments. In the pot experiment the highest S uptake of 153.55 mg pot^{-1} by grain and 297.46 mg pot^{-1} by straw respectively and in the second year of field experiment 17.27 and 21.10 kg ha^{-1} by grain and straw respectively was recorded with this treatment. The other bio-resources application also significantly increased the uptake of S by rice.

The increased uptake of S by rice as a results of the application of various bio-resources was due to the increased supply of nutrients by organic matter during the process of decomposition as well as the creation of conducive soil physical environment under submerged condition for better root proliferation which in turn facilitated higher utilization / absorption of S by rice plants (Minhas and Sood, 1994; Muneshwar Singh *et al.*, 2001). Dekha Medhi *et al.* (1986) reported higher S uptake by rice over the control by the addition of *Sesbania rostrata* / FYM because these sources increased the availability of S in the soil solution.

Table 1. Effect of bio-resource treatments on the secondary nutrients content (%) of rice in the pot experiment

Treat-ments	S				Ca				Mg			
	AT	PI	Grain	Straw	AT	PI	Grain	Straw	AT	PI	Grain	Straw
T_1	0.23	0.21	0.12	0.14	0.52	0.45	0.25	0.32	0.30	0.27	0.21	0.20
T_2	0.28	0.26	0.18	0.21	0.67	0.28	0.30	0.40	0.37	0.30	0.26	0.25
T_3	0.25	0.22	0.13	0.16	0.60	0.50	0.26	0.37	0.34	0.28	0.23	0.22
T_4	0.25	0.24	0.16	0.17	0.61	0.53	0.28	0.38	0.35	0.29	0.24	0.23
T_5	0.24	0.23	0.14	0.13	0.58	0.48	0.26	0.35	0.32	0.28	0.22	0.22
T_6	0.33	0.28	0.21	0.25	0.72	0.62	0.40	0.45	0.42	0.36	0.30	0.27
T_7	0.31	0.27	0.20	0.24	0.69	0.59	0.37	0.43	0.39	0.33	0.29	0.25
T_8	0.29	0.27	0.19	0.22	0.68	0.55	0.35	0.39	0.38	0.31	0.27	0.24
T_9	0.26	0.25	0.19	0.20	0.64	0.54	0.31	0.37	0.36	0.30	0.25	0.24
SEd	0.02	0.02	0.01	0.02	0.01	0.01	0.01	0.02	0.02	0.02	0.01	0.02
CD (p=0.05%)	0.03	0.03	0.02	0.03	0.02	0.02	0.02	0.04	0.04	0.03	0.02	0.04

T_1 - Control (100% NPK), T_2 - 75% N + 25% N through *Sesbania rostrata* +PK, T_3 - 75% N + *Azospirillum* + PK, T_4 - 75% P + 25% P through enriched Farm yard manure + NK, T_5 - 75% P + *Phosphobacterium* + NK, T_6 - 75% N + 25% N through *Sesbania rostrata* + 75% P + 25% P through enriched Farm yard manure + K, T_7 - 75% N + 25% N through *Sesbania rostrata* + 75% P + *Phosphobacterium* + K, T_8 - 75% N + *Azospirillum* + 75% P + 25% P through enriched Farm yard manure + K, T_9 - 75% N + *Azospirillum* + 75% P + *Phosphobacterium* + K

Table 2. Effect of bio-resource treatments on the uptake of secondary nutrients (mg pot^{-1}) by rice in the pot experiment

Treatments	S				Ca				Mg			
	AT	PI	Grain	Straw	AT	PI	Grain	Straw	AT	PI	Grain	Straw
T_1	66.45	69.59	55.73	118.20	83.73	286.82	139.65	295.22	83.73	168.72	111.82	160.52
T_2	86.91	94.70	104.40	182.74	165.00	444.60	215.32	392.86	135.00	273.60	163.17	201.00
T_3	78.99	87.01	79.53	145.60	103.00	342.47	165.17	325.44	96.94	199.77	134.56	179.85
T_4	78.78	86.94	94.92	175.00	101.30	338.94	189.85	351.80	95.34	197.72	151.92	202.45
T_5	79.14	87.58	77.75	142.26	112.50	365.38	167.38	309.65	106.25	217.25	131.53	167.34
T_6	103.90	112.87	153.55	297.46	225.28	531.58	305.72	559.00	177.01	350.78	249.57	297.76
T_7	94.89	103.43	136.00	245.12	213.22	505.98	280.00	481.56	165.83	316.00	219.00	257.35
T_8	92.74	98.71	121.50	234.65	199.80	494.21	250.56	447.25	153.84	294.23	200.00	244.83
T_9	87.83	96.12	116.75	198.97	187.44	451.14	240.10	416.56	148.39	284.93	185.72	208.30
SEd	2.76	3.37	3.45	5.62	6.01	10.90	8.67	6.50	4.86	8.86	4.75	4.89
CD (p=0 .05%)	5.52	6.74	6.89	11.24	12.02	21.65	17.33	13.00	9.72	17.71	9.50	9.77

T_1 - Control (100% NPK), T_2 - 75% N + 25% N through *Sesbania rostrata* +PK, T_3 - 75% N + *Azospirillum* + PK, T_4 - 75% P + 25% P through enriched Farm yard manure + NK, T_5 - 75% P + *Phosphobacterium* + NK, T_6 - 75% N + 25% N through *Sesbania rostrata* + 75% P + 25% P through enriched Farm yard manure + K, T_7 - 75% N + 25% N through *Sesbania rostrata* + 75% P + *Phosphobacterium* + K, T_8 - 75% N + *Azospirillum* + 75% P + 25% P through enriched Farm yard manure + K, T_9 - 75% N + *Azospirillum* + 75% P + *Phosphobacterium* + K

Calcium

The importance of Ca nutrition in salt affected soils is well recognized as it plays important role in ameliorating the soil and it minimize the salt injury to the plants (Gupta and Abrol, 1990). Some important plant functions are controlled by changes in very small physiologically active pools of Ca^{2+} within the cytoplasm (McLaughlin and Wimmer, 1999) and thus Ca nutrition in salt affected soils has a vital role for efficient crop production in these soils.

The increased content and uptake of Ca by rice due to the imposition of bio-resources treatments was well evidenced in the present study. In the field experiments, the positive influence of added bio-resources in enhancing the Ca content and uptake by rice was more pronounced in the second year than in the previous year of rice cultivation (Table 3-6).

In the field experiments, the highest Ca content of 0.29 per cent in grain and 0.48 per cent in straw of the first crop and 0.43 per cent in grain and 0.50 per cent in the straw of the second crop was recorded with green manure and enriched FYM together with 75% NP. This was followed by the addition of *Sesbania rostrata* + *Phosphobacterium*. The effect of other bio-resource treatments was also favorable in this regard.

As like the content, the uptake of Ca by rice was also the highest with the treatment, 75% NP + *Sesbania rostrata* + enriched FYM. A Ca uptake of 305.72 and 559.00 mg pot^{-1} in grain and straw respectively was recorded with this treatment. In the field experiments, this treatment recorded the highest Ca uptake of 20.12 and 32.23 kg ha^{-1} in the grain and straw (first year) and 24.75 and 35.26 kg ha^{-1} in the grain and straw (second year) respectively.

The increased content of Ca and subsequently its uptake by rice due to the application of various bio-resources might be ascribed to reasons such as:

i) Bio-resource amendments itself containing appreciable amount of Ca

ii) Solubilization of Ca compounds during the decomposition of added bio-resources and

iii) The reduction in the ESP of the soil, which implies that the Ca^{++}: Na^{+} ratio in the saline media got adjusted and favored for Ca uptake by rice (Hu and Schmidhatter, 2005).

Patel and Bhajan Singh (1993), Anand Swarup (1992) and Selvi *et al.* (2003b) also reported the increased uptake of Ca by rice due to the application of various bio-resources.

Table 3. Effect of bio-resource treatments on the content of secondary nutrients (%) of rice in the field experiment-I

Treat-ments	S				Ca				Mg			
	AT	PI	Grain	Straw	AT	PI	Grain	Straw	AT	PI	Grain	Straw
T_1	0.25	0.21	0.11	0.15	0.51	0.47	0.28	0.35	0.32	0.28	0.21	0.19
T_2	0.34	0.30	0.20	0.28	0.74	0.65	0.39	0.48	0.45	0.35	0.32	0.28
T_3	0.31	0.28	0.18	0.25	0.67	0.61	0.35	0.45	0.42	0.33	0.28	0.25
T_4	0.28	0.25	0.15	0.23	0.62	0.57	0.31	0.39	0.38	0.32	0.25	0.23
T_5	0.25	0.23	0.13	0.18	0.59	0.52	0.28	0.35	0.35	0.30	0.23	0.20
SEd	0.01	0.02	0.02	0.01	0.03	0.02	0.01	0.02	0.03	0.01	0.02	0.02
CD (p=0 .05%)	0.03	0.04	0.04	0.02	0.05	0.04	0.02	0.04	0.05	0.02	0.03	0.04

T_1- Control (100% NPK), T_2- 75% NP + 25% N through *Sesbania rostrata* + 25% P through Super Phosphate enriched FYM + K, T_3- 75% NP + 25% N through *Sesbania rostrata* + *Phosphobacterium* + K, T_4- 75% NP + *Azospirillum* + 25% P through Super Phosphate enriched FYM + K, T_5- 75% NP + *Azospirillum* + *Phosphobacterium* + K

Table 4. Effect of bio-resource treatments on the content of secondary nutrients (%) of rice in the field experiment-II

Treat-ments	S				Ca				Mg			
	AT	PI	Grain	Straw	AT	PI	Grain	Straw	AT	PI	Grain	Straw
T_1	0.24	0.20	0.11	0.14	0.49	0.45	0.27	0.34	0.30	0.26	0.20	0.19
T_2	0.37	0.32	0.23	0.30	0.77	0.68	0.43	0.50	0.52	0.38	0.36	0.31
T_3	0.32	0.29	0.20	0.27	0.69	0.63	0.37	0.47	0.46	0.36	0.31	0.27
T_4	0.30	0.27	0.18	0.23	0.65	0.59	0.34	0.41	0.42	0.33	0.26	0.23
T_5	0.26	0.24	0.15	0.19	0.61	0.55	0.31	0.37	0.40	0.31	0.24	0.21
SEd	0.02	0.01	0.01	0.02	0.02	0.02	0.02	0.02	0.03	0.01	0.02	0.02
CD (p=0 .05%)	0.03	0.02	0.02	0.03	0.04	0.04	0.03	0.03	0.05	0.02	0.04	0.03

Table 5. Effect of bio-resource treatments on the uptake of secondary nutrients (kg ha^{-1}) **by** rice in the field experiment-I

Treat-ments	S				Ca				Mg			
	AT	PI	Grain	Straw	AT	PI	Grain	Straw	AT	PI	Grain	Straw
T_1	7.28	10.05	4.20	7.83	14.29	22.39	10.72	18.19	9.40	13.52	8.04	9.87
T_2	18.03	21.87	10.83	18.57	39.26	46.50	20.12	32.23	23.48	25.58	16.45	18.55
T_3	14.97	19.04	8.65	16.62	32.50	41.68	16.85	29.85	20.60	22.47	13.50	16.62
T_4	12.39	16.00	6.83	13.25	30.04	36.61	14.00	24.50	16.24	20.53	11.32	14.45
T_5	8.92	12.71	5.55	10.67	21.10	28.82	12.76	22.44	12.37	16.65	9.80	11.82
SEd	0.51	0.68	0.21	0.49	1.34	1.41	0.67	0.99	0.84	0.97	0.57	0.63
CD (p=0 .05%)	1.02	1.35	0.41	0.98	2.67	2.81	1.33	1.97	1.68	1.93	1.13	1.24

Table 6. Effect of bio-resource treatments on the uptake of secondary nutrients (kg ha^{-1}) by rice in the field experiment-II

Treat-ments	S				Ca				Mg			
	AT	PI	Grain	Straw	AT	PI	Grain	Straw	AT	PI	Grain	Straw
T_1	6.76	10.19	4.48	7.64	13.08	21.53	9.64	16.47	8.46	12.87	7.45	9.82
T_2	21.81	25.15	17.27	21.10	45.13	53.16	24.75	35.26	30.55	31.48	18.86	21.45
T_3	16.70	21.46	14.08	18.68	36.11	45.43	18.15	31.82	24.08	26.41	15.42	18.68
T_4	13.72	17.21	12.66	14.25	29.36	38.30	15.96	25.46	18.53	21.84	11.76	14.28
T_5	9.59	13.40	10.11	10.46	22.42	31.29	13.19	21.37	15.50	17.75	10.10	12.62
SEd	0.95	1.44	0.86	0.95	1.44	1.62	0.68	1.49	1.29	1.04	0.64	0.88
CD (p=0 .05%)	1.89	2.88	1.72	1.89	2.87	3.23	1.36	2.97	2.58	2.07	1.28	1.76

Magnesium

The increased content and uptake of Mg by rice at all the critical stages due to the application of different bio-resources was well evidenced in the present investigation.

The highest Mg content in plant samples at critical stages and also in grain and straw was recorded with the application of 75% NP + green manure + enriched FYM (Table 3&4). The Mg uptake in the pot as well as in the field experiments also recorded significantly highest with this same treatment. A significantly highest Mg uptake of 18.86 and 21.46 kg ha^{-1} in grain and straw of second year field experiment was recorded with this treatment (Table 6).

The increased Mg content and uptake by rice in the pot and field experiments was because of the decomposition process of added biomaterials in the soil.

During the decomposition process, a number of organic acids released thus solubilized the $MgCO_3$ in the soil matrix (Mohanty and Patnaik, 1997; Selvi *et al.*, 2003b). In addition to this, the inherent composition of Mg in the organics also improved the Mg nutrition of rice. Patel and Bhajan Singh (1993) also reported an increased uptake of Mg by rice due to the application of various organic materials.

The content and uptake of micronutrients (Fe, Mn, Cu and Zn) by rice in both the field experiments clearly indicated the usefulness of application of various bio-resources in improving the micronutrient nutrition of rice in coastal salt affected soils. All the bio-resources significantly increased the content of micronutrients as well as their uptake by rice at all the critical stages.

In the first year, the content of micronutrients in rice with the application of various bio-resources increased in the range of 40.41 to 49.82 mg kg^{-1} of Fe, 30.50 to 36.75 mg kg^{-1} of Mn, 3.98 to 4.80 mg kg^{-1} of Cu and 15.10 to 18.26 mg kg^{-1} of Zn in rice grain and 346.19 to 382.21 mg kg^{-1} of Fe, 296.85 to 315.25 mg kg^{-1} of Mn, 4.10 to 5.05 mg kg^{-1} of Cu and 22.28 to 27.23 mg kg^{-1} of Zn in rice straw respectively as against the control which registered the lowest content of all these micronutrients.

In the second year of rice cultivation, a further increase in the content of micronutrients was observed. The increased content of various micronutrients in the bio-resources treated plots ranged from 41.24 to 56.36 mg kg^{-1} in grain and 354.28 to 497.50 mg kg^{-1} in straw (Fe), 31.82 to 41.00 mg kg^{-1} in grain and 301.72 to 329.65 mg kg^{-1} in straw (Mn), 4.05 to 4.92 mg kg^{-1} in grain and 4.23 to 5.19 mg kg^{-1} in straw (Cu) and 15.15 to 19.73 mg kg^{-1} in grain and 23.47 to 31.15 mg kg^{-1} in straw (Zn). In tune with the content, the uptake of micronutrients also showed an increase with various bio-resources. Though all the bio-resources evaluated exhibited their significant influence, the application of 75% NP + 25% N substituted through *Sesbania rostrata* + 25% P substituted through enriched FYM excelled in increasing the content and uptake of micronutrients by rice in all the experiments. In the second year, it registered the highest uptake of Fe (216.20 g ha^{-1} in grain and 2988.36 g ha^{-1} in straw), Mn (216.20 g ha^{-1} in grain and 2388.40 g ha^{-1} in straw), Cu (27.08 g ha^{-1} in grain and 38.42 g ha^{-1} in straw) and Zn (102.03 g ha^{-1} in grain and 223.68 g ha^{-1} in straw) respectively.

The soil application of bio-resources especially the green manure and enriched FYM was associated with relatively greater content of Fe, Mn, Cu and Zn by rice at all the critical stages in both the years of rice cultivation. This increased content of micronutrients due to the bio-resources application, might be due to the decreased pH and ESP of the coastal salt affected soil and increased the availability of all the micronutrients. Also the submerged condition of the soil favored the microbial and chemical reduction of these cationic elements for their greater availability (Ponnamperuma, 1972; Anand Swarup, 1991). Furthermore, the bio-resources used in the present investigation contained varied amounts of micronutrients (Table 2) and in turn, made themselves available for the absorption by rice plants. The bio-resources, during decomposition, increased the organic matter content of the soil that contains chelating ligands such as humic and fulvic fractions, which have been observed to function as a sink for the micronutrient cations. These agents have myriad of functional groups, which chelated the micronutrient cations (Decock, 1955; Baniaus *et al.*, 1997) thereby preventing their precipitation in soil even at high pH and increasing their availability. The chelates have made available the various metallic cations near the root surface (Mirchandani, 1992) and facilitated increased absorption by plants. Also the chelates act as carriers of these cations to different parts of the plant resulting in increased content and uptake of these micronutrients. The bio-resources treated plots had higher biomass production so that the uptake of micronutrients increased over the control treatment since the uptake is a function of biomass production. The statistical analysis also indicated the positive correlation between the organic carbon content of the soil and uptake of Fe (r = 0.981* and 0.965** in the first and second year respectively), Mn (r = 0.977** and 0.963** in the first and second year respectively) and Zn (r = 0.980** and 0.963** in the first and second year respectively) by rice.

A vast array of literatures is available for the beneficial influence of bio-resources on the micronutrient nutrition of rice (Anand Swarup, 1991; Selvi and Augustine Selvaseelan, 1991; Sahu *et al.*, 1994; Savithiri *et al.*, 1999; Duhan and Mahendra Singh, 2002).

Table 47. Effect of bio-resource treatments on the content of micronutrients (mg kg^{-1}) at different critical stages of rice in the field experiment- I

Treatments	Fe				Mn				Cu				Zn			
	AT	PI	Grain	Straw	AT	PI	Grain	Straw	AT	PI	Grain	Straw	AT	PI	Grain	Straw
T_1	427.73	341.36	37.34	325.32	323.25	280.55	28.00	288.45	5.85	4.85	3.55	3.93	32.48	26.45	14.70	20.40
T_2	466.05	382.53	49.82	382.21	351.24	301.23	36.75	315.25	6.92	5.70	4.80	5.05	41.40	34.78	18.20	27.23
T_3	459.00	377.71	45.21	373.65	345.82	295.61	34.21	306.18	6.50	5.40	4.48	4.85	39.53	30.48	16.55	25.85
T_4	443.63	355.25	42.00	360.81	337.32	291.11	33.23	301.52	6.20	5.22	4.13	4.53	35.10	28.55	15.47	25.35
T_5	437.52	348.32	40.14	346.19	330.16	285.45	30.50	296.85	5.95	4.73	3.98	4.10	33.28	27.83	15.10	22.28
SEd	3.74	4.69	1.75	4.40	3.74	3.54	0.88	4.61	0.28	0.26	0.22	0.31	1.50	2.32	0.95	1.69
CD (p=0.05%)	7.53	9.42	3.50	8.84	7.52	7.12	1.76	9.21	0.57	0.53	0.45	0.63	3.00	4.66	1.91	3.39

Table. 48. Effect of bio-resource treatments on the content of micronutrients (mg kg^{-1}) at different critical stages of rice in the field experiment- II

Treatments	Fe				Mn				Cu				Zn			
	AT	PI	Grain	Straw	AT	PI	Grain	Straw	AT	PI	Grain	Straw	AT	PI	Grain	Straw
T_1	419.27	336.00	35.50	318.25	315.50	269.95	26.50	276.00	5.79	4.75	3.50	3.89	30.56	28.50	14.25	19.75
T_2	474.15	388.56	56.36	497.50	370.35	310.68	41.00	329.65	7.05	5.95	4.92	5.19	45.15	38.16	19.73	31.15
T_3	465.35	373.20	49.20	382.32	358.26	301.25	37.46	315.21	6.75	5.53	4.60	4.92	41.75	33.50	17.56	27.58
T_4	452.12	361.57	44.16	370.17	344.17	296.00	34.56	308.57	6.35	5.35	4.23	4.67	37.50	30.65	15.98	26.10
T_5	444.20	355.21	41.24	354.28	335.20	289.24	31.82	301.72	6.00	4.89	4.05	4.23	34.27	29.38	15.15	23.47
SEd	4.89	7.14	1.98	9.07	5.41	4.08	1.43	6.59	0.13	0.10	0.09	0.11	1.59	1.08	0.55	0.96
CD (p=0.05%)	9.78	14.27	3.96	18.13	10.82	8.16	2.86	13.18	0.26	0.19	0.17	0.21	3.17	2.16	1.09	1.92

Table.49. Effect of bio-resource treatments on the uptake of micronutrients (g ha^{-1}) at different critical stages of rice in the field experiment- I

Treatments	Fe				Mn				Cu				Zn			
	AT	PI	Grain	Straw	AT	PI	Grain	Straw	AT	PI	Grain	Straw	AT	PI	Grain	Straw
T_1	1247.55	1625.80	145.28	1693.82	944.72	1334.43	107.00	1503.25	17.36	23.12	13.72	20.27	94.78	126.60	54.82	106.63
T_2	2485.29	2792.57	254.34	2627.76	1872.40	2200.56	200.74	2164.90	36.00	41.26	24.70	34.54	220.02	254.37	93.71	188.23
T_3	2230.44	2575.00	218.15	2486.14	1677.20	2014.48	169.45	2042.89	31.32	36.65	21.68	32.03	192.34	208.04	82.98	172.55
T_4	1967.36	2284.78	192.54	2259.20	1495.76	1874.58	146.21	1895.37	27.64	33.30	18.97	28.56	155.70	183.68	71.83	160.25
T_5	1566.56	1935.80	170.71	2058.22	1183.22	1584.50	132.31	1755.96	21.80	26.23	16.50	24.07	119.26	154.60	65.36	131.10
SEd	68.62	76.30	20.03	95.21	63.30	74.48	7.75	71.65	1.16	1.27	0.90	0.21	7.59	9.33	3.75	8.67
CD (p=0 .05%)	137.23	152.59	40.26	191.37	126.50	148.96	15.66	144.02	2.31	2.53	1.82	0.42	15.17	18.65	7.49	17.43

Table.50. Effect of bio-resource treatments on the uptake of micronutrients (g ha^{-1}) at different critical stages of rice in the field experiment-II

Treatments	Fe				Mn				Cu				Zn			
	AT	PI	Grain	Straw	AT	PI	Grain	Straw	AT	PI	Grain	Straw	AT	PI	Grain	Straw
T_1	1184.65	1598.55	127.28	1593.28	873.24	1262.82	94.62	1396.75	15.05	22.87	13.03	19.29	84.17	131.17	51.69	100.54
T_2	2759.60	3029.56	307.20	2988.36	2191.70	2458.10	216.20	2388.40	41.23	47.29	27.08	38.12	266.75	300.78	102.03	223.68
T_3	2414.83	2748.46	241.15	2645.18	1885.28	2186.00	174.80	2188.54	35.02	40.74	23.76	35.17	215.06	242.29	89.56	192.04
T_4	1985.71	2335.96	206.20	2419.77	1542.08	1945.40	152.49	1992.37	28.25	35.24	19.27	32.25	168.38	201.10	74.88	165.66
T_5	1309.42	2049.14	177.37	2141.50	1260.49	1668.24	128.43	1890.85	22.69	28.42	17.48	30.66	128.36	168.50	65.95	141.90
SEd	120.74	128.88	13.26	102.28	114.94	129.05	9.61	94.79	2.99	3.08	1.05	0.84	13.74	18.27	3.86	6.79
CD (p=0 .05%)	241.48	257.76	26.51	204.55	229.88	258.10	19.22	189.58	5.98	6.16	2.09	1.66	27.48	36.53	7.71	13.58

Selected Bibliography

Acharya, C.L., S.K. Bishnoi & H.S.Yaduvanshi (1988). Effect of long term application of fertilizers & organic and inorganic amendments under continuous cropping on soil physical and chemical properties in an alfisol. Indian J. Agric. Sci., 58: 509-516.

Adane A., Gebrekidan H., Kibret K. Effects of treatment application rates (fym and gypsum) on selected chemical properties of saline sodic soils under water limited condition in eastern lowlands Ethiopia. Forest Res Eng Int J. 2019; 3(3): 106113. DOI: 10.15406/freij.2019.03.00086

Aggelider, S.M. & P.A. Londra (2000). Effect of composts produced from town wastes & sewage sludge on the soil physical properties. Bioresour. Technol., 71: 253-259.

Ahmed, K., Qadir, G., Jami, A.R., Nawaz, M.Q., Rehim, A., Jabran, K., Hussain, M. Gypsum and farm manure application with chiseling improve soil properties and performance of fodder beet under saline-sodic conditions. Int. J. Agric. Biol., 2015 17: 1225-1230.

Aich, A.C., R.Mandal and H.M.Ahmed (1996). Impact of lime, gypsum and organic matter on rice grown in saline soils of Bangladesh. J. Indian Soc. Coastal Agric. Res., 14(1&2): 53- 57.

Allen, O.N. (1953). Experiments in soil bacteriology. Burges Publishers, Minnapolis, Minn., U.S.A., pp. 69-70.

Amooaghaie, R., A.Mostajeran and E.Emtiazi (2002). The effect of compatible and incompatible Azospirillum brasilense strains on proton efflux of wheat roots-Azospirillum and proton efflux of wheat root. Plant Soil, 243: 155-160.

Anand Swarup (1988). Influence of organic matter and flooding on the chemical and electrochemical properties of sodic soil and rice growth. Plant Soil, 106: 135-141.

Anand Swarup (1991). Long- term effect of green manuring (Sesbania aculeata) on soil properties and sustainability of rice and wheat yield on a sodic soil. J. Indian Soc. Soil Sci., 39: 777-780.

Anand Swarup (1992). Effect of organic amendments on the nutrition and yield of wetland rice and sodic soil reclamation. J. Indian Soc. Soil Sci., 40: 816-822.

Anderson,G. (1980). Assessing organic phosphorus in soils. In: The role of phosphorus in agriculture. Khasawneh, F.E., Sample,E.C. and Kamprath, E.J. (eds). Am. Soc. Agron, Madison, Wis., pp. 411-431.

Angers, D.A., N. Bissonnette, A.Legere and N. Samson (1993). Microbial and biochemical changes induced by rotation and tillage in a soil under barley production. Can. J. Soil Sci., 73: 39-50.

Anil Kumar Saxena, Suneeta Singh, Jogendra Kumar (2020). Response of scheduled application of zn and fym on yield and micronutrient uptake under rice wheat rotation.International Journal of Ecology and Environmental Sciences 2(4): 215-221.

Arden- Clarke, C. and R.D.Hodges (1988). The environmental effects of conventional and organic / biological-farming systems II: Soil ecology, soil fertility and nutrient cycles. Biol. Agric. Hortic., 5: 223-285.

Azam, F. (1990). Comparative effects of organic and inorganic nitrogen sources applied to a flooded soil on rice yield and availability of nitrogen. Plant Soil, 125: 255-262.

Azmal, A.K.N, T. Matumoto, H. Shindo and M.Nishiyama (1996). Mineralization and microbial biomass formation in upland soil amended with some tropical plant residues at different temperatures. Soil Sci. Pl. Nutr., 42: 463-473.

Babhulkar, P.S., R.M. Wandile, W.P. Badole and S.S. Balpande (2000). Residual effect of long term application of FYM and fertilizers on soil properties (vertisol) and yield of soybean. J. Indian Soc. Soil Sci., 48(1): 89-92.

Babu, S., R.Marimuthu, V.Manivannan and S. Rameshkumar (2001). Effect of organic and inorganic manures on growth and yield of rice. Agric. Sci. Digest, 21(4): 232-234.

Babu, R. and V.C. Reddy (2000). Effect of nutrient sources on growth and yield of direct seeded rice (Oryza sativa L.). Crop res., 19: 189-193.

Babu MVS, Reddy C.M., Subramanyam A., Balaguravaiah D. (2007). Effect of integrated use of organic and inorganic fertilizers on soil properties and yield of sugarcane. Journal of the Indian Society of Soil Science, 55: 161-166.

Badanur, V.P. and S.T. Tolanur (1992). Integrated nutrient management for sorghum and safflower in vertisols. In: Proceedings of International Symposium 'Nutrient Management for Sustained Productivity'. Vol. II, Department of Soils, Punjab Agricultural University, Ludhiana, India. pp. 144-145.

Bagayoko M. (2012). Effects of plant density, organic matter and nitrogen rates on rice yields in the system of rice intensification (SRI) in the "Office Du Niger" in mali. ARPN Journal of Agriculture and Biological Science, 7(8): 620-632.

Baishya, L.K., Rathore, S.S., Dharmendra Singh, Dibyendu Sarkar and B.C. Deka (2015). Effect of Integrated Nutrient Management on rice productivity, profitability and soil fertility. Annals of Plant and Soil Research 17(1): 86-90.

Bajpai, R.K., S.K. Upadhyay, B.S. Joshi and R.S.Tripathi (2002). Productivity and economics of rice (Oryza sativa)-wheat (Triticum aestivum) cropping system under integrated nutrient supply system. Indian J. Agron., 47(1): 20-25.

Balasubramaniyam, P. and S.P.Palaniappan (1990). Studies on the effect of green manuring and nitrogen application in rice- moong cropping system. Indian J. Agron., 35(3): 297-298.

Balasubramaniyan, R. and V.Veerabadran (1997). Substitution of inorganic fertilizer through local green leaf manure and Azozpirillum in rice. Indian J. Agron., 42(2): 285-287.

Ballestero, T.P. & E.M. Douglas (1996). Comparison between the nitrogen fixers from composting farm wastes & composting yard wastes. Trans ASAE, 39: 1709-1715.

Bandyopadhyay, B.K. and D.L.N. Rao (2001). Integrated plant nutrient management in saline soils. J. Indian Soc. Coastal Agric. Res., 19 (1&2): 35-58.

Bandyopadhyay, B.K., A.K. Bandyopadhyay and G.P.Bargava (1985). Characterization of soil potassium and quantity intensity relationship of potassium in some coastal soils. J. Indian Soc. Soil Sci., 33: 548-554.

Bandyopadhyay, B.K., D.Burman and A.Mujumdar (2004). On-farm study on integrated nutrient management for coastal saline soils under rice based cropping system. J. Indian Coastal Agric. Res., 22(1&2): 214-218.

Bandyopadhyay, B.K. and A.K.Bandyopadhyay (1984). Effect of application of farmyard manure on soil properties and yield of crops in coastal saline soils. J. Indian Soc. Coastal Agric. Res., 2: 63-70.

Baniaus, G., P.Securon and G.R.Fithugh (1997). Concentrations and activities of ions in solution displaced from basaltic soils. J. Soil Sci., 28: 454-461.

Baskar, K. (2003). Effect of integrated use of inorganic fertilizers and FYM or green leaf manure on uptake and nutrient use efficiency of rice- rice system on an inceptisol. J. Indian Soc. Soil Sci., 51(1): 47-51.

Basumatary, A. and Talukdar, M.C. (1998). Long term effect of integrated nutrient supply on soil properties in an inceptisol of Assam. Oryza, 35: 43-46.

Bayoumy, M., Khalifa, T. and H. Aboelsoud (2019). Impact of some organic and inorganic amendments on some soil properties and wheat production under saline-sodic soil. J. Soil Sci. Agric. Eng. 2019, 10: 307-313.

Baziramakenga, R. and R.R.Simmard (1998). Low molecular weight aliphatic acid contents of composted manures. J. Environ. Qual., 27: 557-561.

Behera S.K., Singh D. (2009). Effect of 31 years of continuous cropping and fertilizer use on soil properties and uptake of micronutrients by maize (Zea mays)-wheat (Triticum aestivum) system. Indian J. Agri. Sci., 79(4): 264-270.

Bellakki, M.A. and V.P. Badanur (1997). Long term effect of integrated nutrient management on properties of vertisol under dry land agriculture. J. Indian Soc. Soil Sci., 45: 438-442.

Bellakki, M.A., V.P. Badanur and R.A. Setty (1998). Effect of long term integrated nutrient management on some important properties of a vertisol. J. Indian Soc. Soil Sci., 46(2): 176-180.

Bhatt, B.R. Chandra S. Ram, NJAoA Pareek and S. Science (2016). Long term effects of fertilizing and manuring on productivity and soil biological properties under rice (Oryza sativa)- wheat (Triticum sativum) sequence in Mollisols. Arch. Agron. and Soil Sci. 62(8): 109-1122.

Bhandari, A., L. Anil Sood, K.N. Sharma and D.S. Rana (1992). Integrated nutrient management in a rice-wheat system. J. Indian Soc. Soil Sci., 40: 742-747.

Bharadwaj, V. & P.K. Omanwar (1994). Long term effects of continuous rotational croppings and fertilization on crop yields and soil properties II. Effect on EC, pH, organic matter and available nutrients of soils. J. Indian Soc. Soil Sci., 42: 387-393.

Bharambe, P.R., D.K. Shilke, G.S. Jadhav, V.G. Vaishnava and S.R.Oza (2001). Management of salt affected vertisols with sub surface drainage and crop residue incorporation under soybean- wheat cropping system. J. Indian Soc. Soil Sci., 49(1): 24-29.

Bharat Singh and R.P.Pandey (2004). Effect of organic matter and gypsum on electrochemical properties of waterlogged sodic soil. Ann. Pl. Soil Res., 6(2): 220-221.

Bhumbla, D.R. (1972). Reclamation of saline and alkali soils. Indian Fmg., 22: 19-21.

Boddey, R.M. and J. Dobereiner (1988). Nitrogen fixing organisms associated with grasses and cereals: Recent results and perspectives for future research. Plant Soil, 108: 53-65.

Bolton, H., L.F. Elliott, R.I. Papendick and D.F. Bezdicek (1985). Soil microbial biomass and selected enzyme activities: Effect of fertilization and cropping practices. Soil Biol. Biochem., 17: 297-302.

Bonde, A.N., B.G. Karle, M.S. Deshmukh, K.U.Tekale and N.P. Patil (2004). Effect of different organic residues on physico-chemical properties of soil in cotton- soybean intercropping in vertisol. J. Soils and Crops, 14(1): 112-115.

Boparai, B.S., Yadvinder Singh and B.D. Sharma (1992). Effect of green manure (Sesbania aculeata) on physical properties of soil and growth of rice-wheat and maize- wheat cropping system. Int. Agrophys., 6: 95-101.

Bora R., Pandey P.C., Singh D.K., Yadav S.K., Chilwal A. (2018). Assessment of soil fertility status under long term balance fertilizer application on rice (Oryza sativa L.). International Journal of Chemical Studies, 6(5): 1696-1699.

Bremner, J.M., and R.L. Mulvaney (1978). Urease activity in soils. In: Burns, R.G. (ed.), Soil enzymes, Academic press, New York. pp. 149-196.

Broadbent, F.E. (1986). Effects of organic mater on nitrogen and phosphorus supply to plant. In: Chen, Y., Avnimelch, Y. (eds.). The role of organic matter in modern agriculture. Nijhoff, Dor drecht, pp. 13-27.

Brooks, P.C., D.S. Powlson and D.S. Jenkinson (1984). Phosphorus in the soil microbial biomass. Soil. Biol. Biochem., 16: 169-175.

Budhar, M.N. and S.P. Palaniappan (1996). Effect of conjugative and individual application of fertilizer and green manure nitrogen on N uptake and yield of lowland rice. Fertil. News, 41(8): 51-54.

Bulleck III, L.R., M. Brocious, G.K. Evanylo and J.B. Ristaino (2002). Organic and synthetic fertility amendments influence soil microbial, physical and chemical properties on organic and conventional farms. App. Soil Ecol., 19: 147-160.

Burdman, S., G. Dulguerova, O. Okon and E. Jurkevitch (2001). Purification of the major outer membrane protein of Azospirillum brasilense, its affinity to plant roots and its involvement in cell aggregation. Mol. Plant Microbe Interact., 14: 555-561.

Burns, R.C. (1982). Enzyme activity in soil. Location and possible role in microbiology. Soil Biol. Biochem., 14: 107-108.

Casida, L.E., D.A. Klein and Thomas Santoro (1964). Soil dehydrogenase activity. Soil Sci. 98: 371-376.

Celik, I., I. Ortas and S. Kilic (2004). Effects of compost, Mycorhyza manures and fertilizers on some physical properties of a chromoxerert soil. Soil & Tillage Res., 78: 59-67.

Chakrabarti, K., P. Bhattacharya, A. Chakraborti, P. Banik and D.K. Bagchi (1998). Microbial biomass-C and urease activity of soils: Influence of soil properties. Indian Agric., 42(1): 27-36.

Chaudhary, M.L., J.P. Singh and R.P.Narwal (1981). Effect of long-term application of P, N and FYM on soil chemical properties. J. Indian Soc. Soil Sci., 29(1): 81-85.

Chen Lizhu (1988). Green manure cultivation and use for rice in China. Proc. Symp. on Sustainable Agriculture. Green manure in rice farming, IRRI, Philippines. pp. 63-70.

Chesnin,F. and C.H.Yien (1951). Turbidimetric determination of available sulphates. Soil Sci. Soc. Am. Proc. 15: 149-151.

Christiansen-Weniger, C. (1998). Ammonium excreting Azospirillum brasilense C_3: An inhabiting induced tumors along stem and roots of rice. Soil Biol. Biochem., 34: 1083-1092.

Chowdhury, M.H. Rahman & S. Bhattacherjee (2014). Effect of integrated nutrient management on nutrient uptake & economics of fertilizer use in rice cv. NERICA 10 K.J. Bangladesh Agril. Univ. 12(2): 273-277, 2014 ISSN 1810-3030 Ghosh*, M.A.H.

Clarson, D., P.P. Ramaswamy and U.S.Sree Ramulu (1984). Influence of amendments on rice in a sodic soil. Madras Agric. J., 71(10): 681-684.

Coharan, W.G. (1950). Estimation of bacterial densities by means of 'Most Probable Numbers'. Biometrics, 6: 105-116.

Cook, B.D. and D.L.Allan (1992). Dissolved organic matter in old-field soils: total amounts as a measurement of available resources for soil mineralization. Soil Biol. Biochem., 24: 585-594.

Crecchio, C., A.Gelsemino, R. Ambrosoli, J.L.Minati and P. Ruggiero (2004). Functional and molecular responses of soil microbial community under differing soil management practices. Soil Biol. Biochem., 36: 1873-1883.

Dalal, R.C. and R.J.Mayer (1986). Long term trends in fertility of soils under continuous cultivation and cereal croppings in southern Queensland: III distribution and kinetics of soil organic carbon in particle size fractions. Aust. J. Soil Res., 24: 293-300.

Debele, T., S. Sharanappa, K. Sudhir and G.M. Sujith (2001). Direct and interactive effects of enriched farm yard manure and nitrogen levels on the productivity and nutrient uptake of maize. Karnataka J. Agric. Sci., 14(4): 894-899.

Decock, P.C. (1955). Influence of humic acids on plant growth. Sci., 121: 473- 474.

De Datta, S.K. and M.S.A.A.A. Kerim (1974). Water and nitrogen economy of rainfed rice as affected by soil puddling. Soil Sci. Soc. Am. Proc., 38: 515-518.

Deepa Thomas and G. James Martin (1999). Integrated nitrogen for wet seeded rice. Madras Agric. J., 86(10-12): 529-532.

Dekha Medhi, B., H.P. Bharatakur and S.N. Bharathakur (1986). Effect of organic and inorganic sources of nitrogen on the nutrients in the soil and soil solution and growth of rice. J. Indian Soc. Soil Sci., 44: 263-266.

Denison, D.A. and R.D. Koehn (1977). Assay of cellulases. Mycologia LXIX: 592.

Devipriya, P. and K. Kumaraswamy (2004). Grain yield of rice (Oryza sativa L.) crop as influenced by sources of phosphorus with and without organic sources. Oryza, 41 (1&2): 24-27.

Dhumsingh (1994). Biofertilizers: an important component of integrated nutrient system. Indian Fmg., 44 (8): 11-13.

Dick, R.P. (1992). A review: Long-term effects of agricultural systems on soil biochemical and microbial parameters. Agricult. Ecosys. Environ., 40: 25-36.

Dikemann, K.H., S.K. De Datta and J.C.G. Ottow (1993). Nitrogen uptake and recovery from urea and green manure in lowland rice measured by ^{15}N and non- isotope techniques. Plant Soil, 148: 91-99.

Dixit, K.G and B.R. Gupta (2000). Effect of farmyard manure, chemical and biofertilizers on yield and quality of rice (Oryza sativa L.) and soil properties. J. Indian Soc. Soil Sci., 48(4): 773-780.

Doran, J.W. and M.R. Zeiss (2000). Soil health and sustainability: Managing the biotic component of soil quality. Appl. Soil Ecol., 15: 3-11.

Drik L. and Susanne K. (2002). Mobile and readily available C and N fractions and relationship to microbial biomass and selected enzyme activities in a sandy soil under different management systems. J. Plant Nutr. Soil Sci., 165: 9-16.

Duhan, B.S., Devendar Kataria, J.P. Singh, M.S. Kunad and S.S. Dahiya (2005). Effect of nitrogen, farmyard manure and metribuzin on nitrogen transformation. Indian Soc. Soil Sci., 53(2): 184-187.

Duhan, R.S. and Mahendra Singh (2002). Effect of green manuring and nitrogen on yield and uptake of micronutrients by rice. J. Indian Soc. Soil Sci., 50(2): 178-180.

Eghball, B., J. Power, J.E. Gilley and J.W. Doran (1997). Nutrient, carbon and mass loss during compost of cattle feedlot manure. J. Envirn. Qual., 26: 189-193.

El- Shakweer, M.H.A., E.A.El- Sayad and M.S.A.Ejes (1998). Soil and plant analysis as a guide for interpretation of the improvement efficiency of organic conditioners added to different soils in Egypt. Commun. Soil Sci. and Plant Analysis, 29: 2067-2088.

Eriksson, K.E.L., R.A. Blanchette and A. Ander (1990). Biodegradation of cellulose. In: Microbial and enzymatic degradation of wood and wood components.

K.E.L.Eriksson. Blanchette, R.A and Ander, P. (eds.), Springer-verlag, New York. pp. 89-180.

Frankenberger, W.T. and W.A. Dick (1983). Relationship between enzyme activities and microbial growth and activity indices in soil. Soil Sci. Soc. Am. J., 47: 945-951.

Frenzy, J.R. (1986). Forms and reactions of organic sulphur compounds in soils. In: Tabatabai, M.A. (ed.): Sulfur in agriculture. Agron. Mohogr., 27, ASA, Madison, WI, pp. 207-232.

Garcia, C., T. Hernandez, F. Costa and B. Ceccanti (1994). Biochemical parameters in soils regenerated by the addition of organic wastes. Waste Management Res., 12: 457-466.

Garcia, C., T. Hernandez, J.A. Pauscal, J.L. Moreno and M. Ross (2000). Microbial activity in soils of SE Spain exposed to degradation processes. In: Garcia,C., Hernandez,M.T. (eds.), Research and perspectives of soil enzymology in Spain. CEBAS- CSIC, Spain, pp. 93-143.

Garcia-Garrido, J.M., I. Garcia Romea and J.A. Ocampo (1992). Cellulase activity in lettuce and onion plants colonized by the vascular-arbuscular mycorrhizal fungus Glomus mosseae. Soil Biol. Biochem., 25(4): 503-504.

Gaunt, J.L., H.U. Neue, K.G. Cassman, D.C. Olk, J.R.M. Arah, C.Witt, J.C.G. Ottow and I.F. Grant (1995). Microbial biomass and organic matter turnover in wetland rice soils. Biol. Fertil. Soils, 19: 333-342.

Gaur, A.C. (1990). Phosphate solubilizing microorganisms as bio fertilizers. Omega Scientific Publishers, New Delhi. pp. 48-53.

Gaur, A.C. and G. Sunita (1999). Phosphate solubilizing microorganisms- an overview. Curr. trends in Life Sci., 23: 151-164.

Gerresten, F.C. (1948). The influence of microorganisms on the phosphorus uptake by plants. Plant and Soil, 1: 51-81.

Gill1, P.K. and C.S. Aulakh (2018). Effect of integrated nitrogen management on NPK uptake in basmati rice (Oryza sativa L.). Journal of Applied and Natural Science 10(1): 258-261.

Gogoi, P., D.B. Phookan and U. Das (2018). Effect of organic inputs and microbial consortium on yield and soil health of Knolkhol (Brassica oleracea L. var. gongylodes) cultivation. Int. J. Curr. Microbiol. App. Sci., 7(4): 3465-3471.

Golchin, P., P. Clarke, J.M. Oades and J.O. Skemstad (1993). The effect of cultivation on the composition of organic matter and structural stability of soils. Aust. J. Soil Res., 33: 975-983.

Gomez, A.K. and A.A. Gomez (1984). Statistical Procedure for agricultural research, 2nd edn. John Wiley and Sons, New York.

Gopalaswamy, G., V. Narasimhan and A.A. Kareem (1989). Response of direct sown rice to Azospirillum lipoferum, IRRN, 14(5): 24.

Goyal, S., Kapoor, K.K. Daljit Singh & M.C. Mundra (1998). Influence of crop rotations on soil microbial biomass and microbial activities. Indian J. Microbiol., 38: 217-220.

Goyal, S., M.M. Mishra, S.S. Dhankar, K.K. Kapoor and R. Batra (1993). Microbial biomass turnover and enzyme activities following the application of farmyard manure to field soils with and without previous long-term applications. Biol. Fertil. Soils, 15: 60-64.

Goyal, S., M.M. Mishra, S.S. Shankar and K. Chander (1994). Relationship of soil microbial biomass and mineralization of nitrogen following incorporation of organic materials to soil. J. Indian Soc. Soil Sci., 42: 474-476.

Gregorith, E.G., M.R. Carter, D.A. Angers, C.M. Montreal and B.H. Ellert (1994). Towards a minimum data set to assess soil organic matter quality in agricultural soils. Can. J. Soil Sci., 74: 367-385.

Gupta, A.P. and R.P. Narwal (1988). Effect of source and levels of nitrogen and sewage sludge on nitrogen transformation into ammoniacal and nitrate forms. J. Indian Soc. Soil Sci., 36: 352-355.

Gupta, R.K. and I.P. Abrol (1990). Salt affected soils: their reclamation and management for crop production. Adv. Soil Sci., 45: 247-255.

Gupta, R.K., B.R. Arora and K.N. Sharma (2003).Effect of urea and manures addition on the mineral nitrogen content of the texturally divergent soils. J. Indian Soc. Soil Sci., 51(2): 203-205.

Gupta, R.P. and C. Dakshinamoorthy (1980). Procedure for physical analysis of soils and collection of agrometeriological data. Division of Agrl. Physics, Indian Agricultural Research Institute, New Delhi.

Hafeez, B., Y. Kanif and M.J. Ajoe A. Saleem (2013). Role of Zinc in plant nutrition-a-review. American J. Experimental Agriculture. 3(2): 374.

Han, X., X. Cong, A. Jennifer, J. Dungait, R. Bol, X. Wang, W. Wu and F. Meng (2018). Straw incorporation increases crop yield and soil organic carbon sequestration but varies under different natural conditions and farming practices in China: a system analysis. Biogeosciences, 15: 1933-1946.

Hayman, D.S. (1975). Phosphorus cycling on soil microorganisms and plant roots. In: Walker, N (ed.) Soil microbiology. Willey and Sons, New York, pp. 67-91.

Hedge, D.M. and B.S. Dwivedi (1994). Crop response to biofertilizers in irrigated areas. Fert. News, 39(4): 19-26.

Haider Iqbal Khan (2018).Appraisal of Biofertilizers in Rice: To Supplement Inorganic Chemical Fertilizer. Rice Science. 25(6): 357-362.

Hemalatha, M., V. Thirumurugan, M. Joseph and R. Balasubramaniyan (2001). Effect of different sources of nitrogen on growth and yield of rice. J. Maharastra Agric. Univ., 25(3): 255-257.

Houben, D. and PJAG Sonnet (2012). Zinc mineral weathering as affected by plants roots. Applied Geochemistry. 27(8): 1587-1592.

Howida Elhabet (2018). Effect of Organic and Inorganic Fertilizers on Rice and Some Nutrients Availability under Different Water Regimes. J Agri Sci Food Res. 9(4): 3-16.

Hu, Y. and U. Schmidhalter (1997). Interactive effect of salinity and macronutrient level on wheat. 2. Composition. J. Plant Nutr., 20: 1169-1182.

Hu, Y. and U. Schmidhalter (2005). Drought and salinity: A comparison of their effects on mineral nutrition of plants. J. Plant Nutr. Soil Sci., 168: 541-549.

Humphries, E.C. (1956). Mineral components and ash analysis. In: Modern methods of plant analysis. Springer Verlag, Berlin, 1: 468-502.

Hundal, H.S. and S.S.Thind (1993). Release of phosphorus in flooded soil amended with green manure. J. Indian Soc. Soil Sci., 41(4): 782-783.

Hwangbo, H., Park, R.D., Kim, Y.W., Rim, Y.S., Park, K.Y., Kim, T.H., Suh, J.S., Kim, K.Y. 2-Ketogluconic acid production and phosphate solubilization by Enterobacter intermedium. Current Microbiology, 2003; 47: 87-92.

Imagawa, M., S. Kewai, T. Kinoshita, H. Myumi and H.Oshima (1989). Accumulation of carbon in soil brought out by application of rice straw and its effect on property and productivity of mineral upland soil. Res. Bull. Aichiken Agric. Res. Centre, 21: 281-288.

Iqbal M., Van Es, H.M., Hansan A., Schindelbeck R.R., Moebivs-Clune B.N. (2014). Soil Health indicators as affected by long term application of farm manure and cropping patterns under semi arid climates. International Journal of Agriculture and Biology, 16: 242-250.

Jackson, M.L. (1973). Soil chemical analysis. Prentice Hall of India Pvt. Ltd., New Delhi.

Jadhav A.D., Talashilkar S.C. and A.G. Powar (1997). Influence of conjunctive use of FYM, vermicompost and urea on growth and nutrient uptake in rice. Journal of Maharashtra Agricultural Universities 22(2): 249-250.

Jahiruddin, M. and M.S. Cresser (1991). Solubility of Zn as affected by Fe, Al and P concentrations in relation to soil pH, J. Indian Soc. Soil Sci., 39(2): 371-373.

Jane J. Kapkiyai, Nancy K. Karanja, Javid N. Qureshi, Paul. C. Smithson and Paul L. Woomer (1999). Soil organic matter and nutrient dynamics in a Kenyan nitisol under long-term fertilizer and organic input management. Soil Biol. Biochem., 31: 1773-1782.

Jayabal, A., S.P. Palaniappan and S. Chellaih (1999). Evaluation of integrated nutrient management techniques in rice. Oryza, 36(3): 263-265.

Jenkinson, D.S., R.H. Fox and J.H. Rayner (1985). Interactions between fertilizer nitrogen and soil nitrogen, the so-called 'priming effect'. J. Soil Sci., 36: 425-444.

Kabirinejad, S., M. Kalbasi, A.H. Khoshgoftarmanesh, M. Hoodaji and M. Afyuni (2014). Effect of incorporation of crops residue into soil on some chemical properties of soil and bioavailability of copper in soil. Int. J. Adv. Biol. Biomedical Res., 2(11): 2819-2824.

Kanivets, I.I., N. Ya Krivich and Z.A. Klugen (1971). The fungus Trichoderma lignorum and its role in improving soil fertility and the yield and quality of agricultural crops. Inoculation with organisms other than Rhizobium, C.A.B, p. 9.

Kathirvelan, P. and V. Vaiyapuri (2002). Integration of organic, inorganic and biofertilizers on growth and yield of transplanted rice. National Seminar on Recent Trends on the use of Humic Substances for Sustainable Agriculture, held at Annamalai University, pp. 40-41.

Katkar, R.N., Kharche, V.K., Sonune, B.A., Wanjari, R.H., and Muneshwar singh (2012). Long term effect of nutrient management on soil quality and sustainable productivity under sorghum-wheat crop sequence in vertisol of Akola, Maharashtra. Agropedology, 22(2): 103-114.

Kaushik, A., R.Nisha, K. Jagjeeta and C.P. Kausik (2005). Impact of long and short term irrigation of a sodic soil with distillery effluent in combination with bio amendments. Bioresour. Technol., 96: 1860-1866.

Kenknight, G. and J.H. Munie (1939). Isolation of phytopathogenic actinomycetes. Phytopathology, 29: 1000-1001.

Khanda, C. and L. Dixit (1995). Effect of zinc and nitrogen fertilizer on summer rice (Oryza sativa). Indian J. Agron., 40(4): 695- 697.

Khanda, C.M., B.K. Mandal and L.M. Garnayak (2005). Productivity and economics of different rice based cropping sequences as influenced by integrated nutrient management. Oryza, 42(1): 48-51.

Khatun, M.; Shuvo, M.A.R.; Salam, M.T.B.R.; Hafizur, S.M. Effect of organic amendments on soil salinity and the growth of maize (Zea mays L.). Plant Sci. Today 2019, 6: 106-111.

Kher, D. (1993). Effect of continuous liming, manuring and cropping on DTPA extractable micronutrients in an alfisol. J. Indian Soc. Soil Sci., 41: 366-367.

Kitila, K., Chala, A., Workina, M. Effect of Gypsum and compost Application in Reclaiming Sodic soils at Small Scale Irrigation Farm in Bora District of East Shewa Zone, Oromia, Ethiopia. Agriways 2020, 08: 28-44.

Kolambe, B.N. and M.L. Patel (1994). Efficient utilization of P by transplanted rice in vertisols. Fert. News, 39(2): 39-45.

Kotha Sammy Reddy, Muneshwar Singh, Anand Swarup, Annagi Subba Rao and Kamlesh Narayan Singh (2002). Sulfur mineralization in two soils amended with organic manures, crop residues and green manures. J. Plant Nutr. Soil Sci., 165: 167-171.

Kremner, R.J. and J.Li. (2003). Developing and suppressive soils through improved soil quality management. Soil Tillage Res., 72: 193-202.

Kshattriya, S., G.D. Sharma and R.R. Mishra (1992). Enzyme activities related to litter decomposition in forests of different age and altitudes in northeast India. Soil Biol. Biochem., 24(3): 265-270.

Kukerja, K., M.M. Mishra, S.S. Dhankar, K.K. Kapoor and A.P. Gupta (1991). Effect of long-term manurial application in microbial biomass. J. Indian Soc. Soil Sci., 39: 681-684.

Kuldip Singh and M.S.Aulahb. 2001. Nitrogen immobilization and mineralization in a sandy loam soil (Typic ustochrepts) amended with various crop residues at various moisture regimes. Indian J. Agric. Sci., 43: 73- 74.

Kulvinder Kaur, Krishnan K. Kapoor and Anand P.Gupta. 2005. Impact of organic manures with and without mineral fertilizers on soil chemical and biological properties under tropical conditions. J. Plant Nutr. Soil Sci., 168: 117- 122.

Kumar,A., D.S.Yadav and A.Kumar. 1995. Use organic manure and fertilizer in rice (Oryza sativa)-wheat (Triticum aestivum) cropping system for sustainability. Indian J. Agric. Sci., 65: 703- 707.

Kumar,S. and K.K.Kapoor. 1995. Soil microbial biomass, dehydrogenase and urease activities and decomposition of Sesbania green manure in alkaline soils. Indian J. Microbiol., 35(1): 39- 42.

Kumar B, Gupta RK, Bhandari AL (2008). Changes after long term application of organic manures and crop residues under rice-wheat system. Journal of Indian Society of Soil Science, 56: 80-85

Kumar V, Singh AP (2010). Long-term effect of green manuring and farmyard manure on yield and soil fertility status in rice-wheat cropping system. Journal of Indian Society of Soil Science, 58: 409-412.

Kumar,S.A., H.Chidanandappa and Babu. 2010. Effect of different sources of zinc on growth, yield and uptake of nutrients by Maize crop (Zea mays L.). 44(1): 92-99.

Kuppuswamy, G. A.R. Lakshmanan and A. Jayabal. 1991. Effect of gypsum enriched biogas sludge and farmyard manure on rice yield. IRRN, 16 (4): 18.

Kuppuswamy, G., A. Jayabal and A.R. Lakshmanan. 1992. Effect of enriched bio digested slurry and FYM on growth and yield of rice. Agric. Sci. Digest, 1 (2): 101- 104.

Kyi Moe., Seinn Moh Moh, Aunj Zaw Htwe, Yoshimori Kajihara and Takeo Yamakawa. 2019. Effects of Integrated Organic and Inorganic Fertilizers on Yield and Growth Parameters of Rice Varieties . Rice Science,26(5):309-318.

Lal, B., J.N.Mishra and A.K.Sarkar. 2000. Effect of plant residues incorporation of specific microbial groups and availability of some plant nutrients in soil. J. Indian Soc. Soil Sci., 48(1): 67- 71.

Latha, M. P.Savithri, R.Indirani and S.Kamaraj. 2001. Influence of zinc enriched organic manure on the yield, dry matter production and zinc uptake of maize. Acta Agronomica Hungarica. 49(3):231-236.

Lavanya, P.G. and T.S.Manickam, 1991. Organic manures and their interactions with inorganic fertilizers on nutrient availability, uptake and yield of ragi crop. Madras Agric. J., 78(5-8): 248- 253.

Laxminarayana, K. 2005. Effect of P solubilizing microorganisms on yield of rice & nutrient availability in an acid soil of Mizoram. J. Indian Soc. Soil Sci., 53(2): 240- 243.

Liang, M., M.Nicolic, Y.Peng, W.Chen and Y.Jiang. 2005. Organic manure stimulates biological activity and barley growth in soil subject to secondary salinization. Soil Biol. Biochem., 37: 1185-1195.

Leo Daniel Amalraj,E., G.Praveen Kumar,S. Mir Hassan Ahmed, R. Abdul and NJOA Kishore.2013. Microbiological analysis of panchakavya, vermicompost and FYM and their effect on plant growth promotion of pigeon pea (Cajanus cajan L.) in India. Geoderma. 3(1):23-29.

Liang,L.C., W.Y, Lee and C.HJIc Hung.2003. Amido phosphate complexes of zinc. Inorganic Chemistry. 42(18):5471-5473.

Lindsay, W.L. 1972. Zinc in soils and plant nutrition. Adv. Agron., 24: 147- 186.

Lindsay, W.L. and W.A.Norwal. 1978. Development of DTPA soil test for Fe, Mn, Zn and Cu. Soil Sci. Soc. Amer. Proc., 42: 421- 428.

Linkens, A.E. and R.L.Sinsabaugh. 1990. Comparison of cellulase activity on decomposing leaves in a hardwood forest and woodland stream. Soil Biol. Biochem., 22(3): 423-425.

Lynch, J.M. and J.M.Whipps. 1990. Substrate flow in the rhizosphere. Plant Soil, 129: 1-10.

Mahapatra, A.K. and R.C.Jee. 1993. Response of lowland rice to green manuring levels and sources of phosphatic fertilizer in coastal Orissa. Indian J. Agron., 38: 374- 377.

Mahapatra, P., M.M.Panda, A.B.Chalam, S.P.Chakravorthi and S.K.Mohanty. 1997. Effect of green manuring and N fertilization on the yield and nitrogen use efficiency of wetland rice. J. Indian Soc. Soil Sci., 45(1): 95- 99.

Maharudrappa, A., C.A.Srinivasamurthy, M.S.Nagaraja, R.J.Siddaramappa and H.S.Anand. 2000. Decomposition rate of litter and nutrient release pattern in a tropical soil. J. Indian Soc. Soil Sci., 48(1): 92- 97.

Maji,B. and B.K.Bandyopadhyay. 1990. Response of rice to soil and foliar application of micronutrients in coastal soils of sunderban, West Bengal. J. Indian Soc. Coastal Agric. Res., 8(1): 390-392.

Malarkodi, R. and P. Singaram. 2003. Effect of integrated phosphorus management on rice yield, grain quality and total uptake of nutrients. Madras Agric. J., 90 (10-12): 661-664.

Malarkodi, R. and P. Singaram. 2004. Relative efficiency of Jhabua rock phosphate with super phosphate, farmyard manure and Phosphobacteria on available nutrients and heavy metal content in soil under rice. Madras Agric. J. 91(1-3): 96-99.

Malik, A., A.Kausik and C.P.Kausik. 2004. Salinization effect on dehydrogenase activity and CO_2 evaluation from soils following phytomass amendments. Proc. Ind. Natl. Sci. Acad. B 61(3): 181- 186.

Malo, A Ghosh, D Dutta and K Murmu. 2018. Effect of inorganic and bio-fertilizers on growth and yield of rice in New Alluvial Zone of West Bengal M Journal of Pharmacognosy and Phytochemistry 2018; 7(1): 576-580.

Mandal, B.K., N.C.Das, Y.V.Singh and R.K.Ghose. 1993. Use of Azolla and other organic materials for rice production. Oryza, 30(1): 54- 59.

Mandal, U.K., Gurcharan Singh, U.S. Victor and K.L. Sharma. 2003. Green manuring: It's effect on soil properties and crop growth under rice- wheat cropping system. Europ. J. Agronomy, 19: 225- 237.

Manna, M.C., A.Swarup, R.H.Wanjari, H.N Ravankar, B.Mishra, M.N.Saha, D.K.Sahi and P.A.Sarap. 2005. Long term effect of fertilizers and manure application on soil organic carbon storage, soil quality and yield sustainability under sub- humid and semi- arid tropical India. Field Crop Res., 93: 264- 280.

Manoj K., Bhatt,Kiran P. Raverkar,Ramesh Chandra,Navneet Pareek,Rini Labanya,Vineet Kumar.2019. Effect of long-term balanced and imbalanced inorganic fertilizer and FYM application on chemical fraction of DTPA-extractable micronutrients and yields under rice–wheat cropping system in mollisols. Soil Use and Management. 36(2):261-273. https://doi.org/10.1111/sum.12560.

María Alcívar , Andrés Zurita-Silva, ID , Marco Sandoval , Cristina Muñoz and Mauricio Schoebitz 2018. Reclamation of Saline–Sodic Soils with Combined Amendments: Impact on Quinoa Performance and Biological Soil Quality. Sustainability.10.3083. doi:10.3390/su10093083

Marinari,S., G.Masciandaro, B.Ceccanti and S.Grego. 2000. Influence of organic and mineral fertilizers on soil biological and mineral properties. Bioresour. Technol., 72: 9- 17.

Martens, D.A. 2001. Nitrogen cycling under different soil management systems. Adv. Agron., 70: 143-192.

Martens,D.A., J.B.Johanson and W.T.Frankenberger Jr. 1992. Production and persistence of soil enzymes with repeated addition of organic residues. Soil Sci., 153: 53- 61.

Martin. J.P. 1950. Use of acid, Rose Bengal and Stepromycin in the plate method for estimating soil fungi. Soil Sci., 69: 225-233.

Masciandaro,G., B.Ceccanti, S.Benidicto and H.Lee. 2001. Humic substances to reduce salt affect on plant germination and growth. Commun. Soil Sci. Plant Analysis, 33: 3-4.

Masciandaro, G., B.Ceccanti, S.Benidicto, H.C.Lee and H.F.Cook. 2004. Enzyme activity and C and N pools in soil following application of mulches. Can. J. Soil Sci., 84: 19-30.

Mathur, G.M. 1997. Effect of long term application of fertilizers and manures on soil properties and yield under cotton-wheat rotation in northwest Rajastan. J. Indian Soc. Soil Sci., 45: 288- 292.

Maurya, P.R. and A.B.Ghosh. 1972. Effect of long term manuring and rotation cropping on fertility status of alluvial calcareous soil. J. Indian Soc. Soil Sci., 20: 31- 45.

Mayalagu, K., S.Jaganathan and R.Vadivelkumar. 1995. Residual effect of organic manures to red sandy loam under rice- fallow cropping system. National Symp. on Organic Farming held at AC & RI, Madurai, Tamilnadu, India, pp: 15.

Mays, D.A., G.L.Terman and J.C.Duggan. 1973. Municipal compost: effects on crop yields and soil properties. J. Environ. Qual., 2: 89-92.

McLatchy, G.P. and K.R.Reddy. 1998. Regulation of organic matter decomposition and nutrient release in a wetland soil. J. Environ. Qual., 27: 1268- 1274.

McLaughlin, S.B. and R.Wimmer. 1999. Calcium physiology terrestrial ecosystem processes. New Phytol., 142: 373- 417.

Medina, A., N.Vassilev, M.M.Alguacil, A.Roldan and R.Azcon. 2004. Increased plant growth, nutrient uptake and soil enzymatic activities in a desertified Mediterranean soil amended with treated residues and inoculated with native mycorrhyzal fungi and a plant growth promoting yeast. Soil Sci., 169(4): 260- 270.

Mengal, K. and E.A.Kirkby 1982. Principles of plant nutrition. International Potash Institute, Berne, Switzerland.

Michalak, I., R. Wilk and K. Chojnacka. 2017. Bio conversion of Baltic Seaweed into organic compost. Waste Biomass Valor., 8: 1885-1895.

Minhas, R.S. and A.Sood. 1994. Effect of inorganics and organics on the yield and nutrient uptake by three crops in a rotation on an alfisol. J. Indian Soc. Soil Sci., 37: 257-260.

Minhas,P.S., D.R.Sharma and Y.P.Singh. 1995. Response of rice and wheat to applied gypsum and farmyard manure on an alkali water irrigated soil. J. Indian Soc. Soil Sci., 43(3): 452- 455.

Mirchandani, T.B. 1992. Use of micronutrient fertilizers with chelating agent-coating agromin on horticultural crops. A scientific study. Micronutr. News, 5(10, 11, 12): 1-12.

Mishra, B. 2005. Managing soil quality for sustainable agriculture. J. Indian Soc. Soil Sci., 53(4): 529- 536.

Mishra, B., Sharma, A., Singh, S.K., Prasad, J., Singh, B.P. Influenced of continuous application of amendments to maize- wheat cropping system on dynamic of soil microbial biomass in alfisol of Jharkhand. Journal of the Indian Society of Soil Science, 2008; 56: 71-75.

Mishra, V.K. and R.B.Sharma. 1997 Influence of integrated nutrient management on soil health and energy requirement of rice based cropping systems. Oryza, 34: 165- 170.

Modak, S.J. and K.N.Chavan. 1999. Response of rice to various levels of zinc and iron with and without FYM in black calcareous soil. Madras Agric. J., 86(4- 6): 206- 209.

Mohana Rao Puli, P R K Prasad, B Srihari Rao and M Jayalakshmi.2017. Effect of Organic and Inorganic Sources of Nutrients on Secondary and Micro Nutrient Uptake by Rice at various Growth Periods. Research Journal of Agricultural Sciences 8(1): 20-24

Mohanty, S.K. and S. Patnaik. 1997. Effect of submergence on the chemical changes in different rice soil. II. Kinetics of K, Ca and Mg. Acta Agronamica Acad. Sci. Hung., 26: 187- 191.

More, S.D. 1994. Effect of farm wastes and organic manures on soil properties, nutrient availability and yield of rice- wheat grown on sodic vertisol. J. Indian Soc. Soil Sci., 42: 253-256.

Mukhrjee, B.K. and R.R.Agarwal. 1951. Review of green manuring in India. ICAR Bull. No.68.

Mulvaney, R.L. and J.M.Bremner. 1981. Control of urea transformations in soils. In: Soil Biochemistry, Chap. 4, Vol. 5. Paul,E.A. and Ladd,J.M. (eds.), Marcel Dekker, New York

Muneshwar Singh, V.P.Singh and K.Sammi Reddy. 2001. Effect of integrated use of fertilizer nitrogen and farmyard manure or green manure on transformation of N, K and S and production of rice- wheat system on a vertisol. J. Indian Soc. Soil Sci., 49(3): 430-435.

Myint, A., Yamakawa, T., Kajihara, Y, T. 2010. Application of different organic and mineral fertilizers on the growth, yield and nutrient accumulation of rice in a Japanese ordinary paddy field Science. World Journal, 5, 47- 54.

Mythili, S., K Natarajan and T.Chiteshwari. 2001. Impact of green manuring on the availability of sulphur and zinc in the rice soils. Madras Agric. J., 88(7-9): 447- 450.

Nagendra Rao,T., B.Seshaiah, R.Prasad, P.Ravindra Babu, V.Sathyanarayana Rao and R.N.Pillai. 1985. Effect of FYM, Phosphorus and zinc application on their availability. Andhra Agric. J., 35(3-4): 215- 217.

Nahar, K., J.Haider and A.J.M.S.Karim. 1995. Residual effect of organic manure and influence of nitrogen fertilizer on soil properties and performance of wheat. Ann. Bangladesh Agric., 5(2): 73- 78.

Nambiar, K.K.M. and I.P.Abrol. 1992. Long term fertilizer experiments in India: an overview. Fertil. News, 34: 11- 26.

Nannipieri, P. 1994. The potential use of soil enzymes as indicators of productivity, sustainability and pollution. In: Pankhrust, C.E., B.M.Doube, V.V.S.R.Gupta, P.R.Grace (eds.) Soil biota, management in sustainable farming systems. CSIRO, Melbourne, pp.238-244.

Nayar, V.K. and I.M.Chhibba. 2000. Effect of green manuring on micronutrient availability in rice- wheat cropping system in northwest India. In: Abrol, I.P., Bronson, K.F., Duksbury, J.M and Gupta, R.M. (eds.), Long term soil fertility experiments in rice-wheat cropping systems. Rice- wheat consortium paper series 6, Rice- wheat consortium for the Indo- Gangatic plains, New Delhi, pp. 68- 72.

Neelam Sharad and T.S.Verma. 2001. Improvement in soil productivity with long-term lantana (Lantana camara. L) addition in rice- wheat cropping- I. Effects on some soil properties and zinc fractions. J. Indian Soc. Soil Sci., 49(3): 462- 466.

Naher,M.S. and A.K. Paul. 2017. Effect of Integrated nutrient Management on nutrient uptake andsustainable grain yield in transplanted Aman rice. SAARC J. Agri., 15(1): 43-53 (2017) D

Nelidov, S.N., L.V.Vasilevo and E.N.Mishustin. 1998. Use of crop residues to increase productivity of rice on ameliorated alkaline soils. Biol. Bull. Academy of the Sci., USSR, 1:31-43.

Neni Marlina, Nuni Gofar, Abdullah Halim Perdana Kusuma Subakti and Abdul Madjid Rahim.2014.Improvement of rice growth and productivity through balance application of inorganic fertilizer and biofertilizer in inceptisol soil of low land swamp area.Agrivita. 36 (1):48-56.

Nuni Gofar, Abdullah Halim Perdana Kusuma Subakti and Abdul Madjid Rahim.2014. Improvement of Rice growth and productivity through balance application of inorganic fertilizer and biofertilizer in inceptisol soil of low land swamp area. Agrivita .36(1): 48-56.

Nyakatawa, E.C., K.C.Reddy and K.R.Sistani. 2001. Tillage, cover cropping and poultry litter effects on selected soil chemical properties. Soil Till. Res., 58: 69-79.

O'Toole, P. 1991. Assay of urease enzyme activity in an ammonium fixing soil. Commun. Soil Sci. Plant Analysis., 22: 213-224.

Oades, J.M. 1984. Soil organic matter and structural stability: Mechanisms and implications for management. Plant Soil, 76: 319-337.

Oberson, A., J.M.Besson, N.Maire and H.Sticher. 1996. Microbiological transformations in soil organic phosphorus transformation in conventional and biological cropping systems. Biol. Fertil. Soils, 21: 138-148.

Olsen, S.R., C.V. Cole, F.S.Watanabe and L.A.Dean. 1954. Estimation of available phosphorus in soils by extraction with sodium bicarbonate. USDA Circ. 939.

Omarhattab, K., K.Natarajan and A.Gopalaswamy. 2000. Effect of organic and inorganic nitrogen combination on rice yield and N uptake. J. Indian Soc. Soil Sci., 48(2): 398-400.

Ondrasek, G.; Rengel, Z. Environmental salinization processes: Detection, implications & solutions. Sci. Total Environ. 2021, 754, 142432.

Palniappan,S.P and K.Natarajan. 1993. Practical aspects of organic matter maintenance in soil. In: Organics in soil health and crop production. P.K.Thampan (ed.) Peeky Tree Crops Development Foundation, Cochin. Pp. 23-41.

Pani, B.K., Sarongi, C.R. and A.K.Behera. 1998. Response of wetland rice to combined application of Phosphobacteria and inorganic phosphate in alfisol. Indian J. Agric. Sci., 68(6): 330-332.

Pandey, N., Verma A.K., Anurag and Tripathi, R.S. (2007).Integrated nutrient management in transplanted hybrid rice (Oryza sativa L.). Journal of Agronomy 52(1):40-42.

Pankhurst,C.E., B.L.Blair, R.C.Magaray, G.R.Stirling, M.J.Bell and A.L.Garside. 2005. Effect of rotations of breaks and organic matter amendments on the capacity of soils to develop biological suppression towards soil organisms associated with yield decline of sugarcane. Appl. Soil Ecol., 28: 271-282.

Paramanik, S.C. and B.S.Mahapatra. 1997. Effect of integrated use of inorganic and organic nitrogen sources on mineralization, uptake and grain yield of rice. Oryza, 34: 181-814.

Parham,J.A., S.P.Deng, W.R.Raun and G.V.Johnson. 2002. Long term cattle manure application in soil. Effect on soil phosphorus levels, microbial biomass C and dehydrogenase and phosphatase activities. Biol. Fertil. Soils, 35: 328- 337.

Patel, K.P. and Bhajan Singh. 1993. Relative efficiency of gypsum, pressmud and pyrites as affected by time of application under percolation conditions in reclamation of alkali soils. Ann. Agric. Res., 14(1): 77- 83.

Patel, M.S., R.G.Patel, N.M.Zalawadia and G.S.Saturia. 1992. Effect of phosphorus, moisture and organic matter on phosphorus availability and its fractions in a clay calcareous soil at different incubation periods. J. Indian Soc. Soil Sci., 40: 705- 710.

Pathak, A.N., D.N.Sharma and M.L.Sharma. 1975. Effect of soil salinity on germination and nutrient uptake of paddy (IR-8) and wheat (K-65) crop. Indian J. Farm Sci., 3: 64-68.

Patiram, Singh KA. (1993). Effect of continuous application of manure and nitrogenous fertilizer on some properties of acid Inceptisol. Journal of Indian Society of Soil Science, 1993; 41:430-433.

Patra, A., V.K. Sharma, T.J. Purakayastha, M. Barman, S. Kumar, D. Chakraborty, A. Kapil Chobhe and D.J. Nath. 2018. Long-term effect of integrated nutrient management on yield and nutrients uptake by rice (Oryza sativa) in acid soil. Indian J. Agric. Sci., 88 (4): 579-583.

Pattanayak, S K, Mishra K .N, Jena M .K and R.K. 2001. Evaluation of green manure crops fertilized with various phosphorus sources and their effect on subsequent rice crop. Journal of the Indian Society of Soil Science 49(2): 285- 291.

Paul, E.A. and F.E.Clark. 1996. Soil microbiology and biochemistry, 2nd Ed. Academic Press, San Diego, CA.

Pauscal, J.A., T.Hernandez, C.Garcia and M.Ayuso. 1998. Enzymic activities in an arid soil amended with urban organic wastes: laboratory experiments. Bioresour. Technol., 64: 131-138.

Perez-Matoes, M., S.Gonzales-Carcedo and M.D.Busto Nuniz. 1988. Extraction of catalase from soil. Soil Sci. Soc. Am. J., 52: 408-411.

Perotti, E.B.R. and A.Pidello. 1999. Effect of Azospirillum brasilense inoculation on urease activity in soil and gamma- sterilized soil. Revista Argentina de Microbiol., 31: 36-41.

Perucci, P. 1990. Effect of the addition of municipal solid waste compost on microbial biomass and enzyme activities in soil. Biol. Fertil. Soils, 10: 221- 226.

Perucci, P. 1992. Enzyme activity and microbial biomass in a field soil amended with municipal refuse. Biol. Fertil. Soils, 14: 54-60.

Piper, C.S. 1966. Soil and plant analysis. Interscience Publishers, New York.

Polara, K.B., M.S.Patel, N.K.kalyanasunderam and J.V.Polara. 2002. Effect of gypsum and FYM application at varying leaching levels on yield and nutrient uptake by wheat in saline sodic soils. J. Indian Soc. Coastal Agric. Res., 20(2): 77- 84.

Ponnemperuma, F.N. 1972. The chemistry of submerged soils. Adv. Agron., 24: 29-96.

Ponnemperuma, F.N. 1977. Physiochemical properties of submerged soils in relation to fertility. IRRI Res. Series 5: 11-12.

Poongothai,S., S.Mani and Rani Perumal. 1995. Effectiveness of zinc enriched organic manures to rice-groundnut cropping system in typic haplustalf. National symposium on Organic farming. Oct. 27-28. AC&RI., Madurai. pp.57-58.

Powar, S.L., S.N.Jadhav and L.Sannanse. 2004. A long- term effect of integrated nutrient management on soil properties and crop yield in rice- rice cropping system in coastal ecoregion of Maharastra. J. Indian Soc. Coastal Agric. Res., 22 (1&2): 250-252.

Powar, S.L.2002. Effect of levels of fertilizers and green manure on rice in medium black soil. J. Indian Soc. Coastal Agric. Res., 20(2): 119- 122.

Prabhakar, A.S., S.V.Patil and K.Krishnamoorthy. 1972. Influence of organic manures, ammoniacal and nitrate nitrogen on the availability of soil and applied phosphorus. J. Indian Soc. Soil Sci., 20(4): 413- 415.

Prakash,Y.S., P.B.S.Bhadoria and Amitava Rakshit. 2002. Comparative efficacy of organic manures on the changes in soil properties and nutrient availability in an alfisol. J. Indian Soc. Soil Sci., 50(2): 219- 221.

Prasad, B. 1994. Integrated nutrient management for sustainable agriculture. Fertil. News, 39: 19- 25.

Prasad, B. and R.P.Singh. 1981. Accumulation and decline of available nutrients with long-term use of fertilizers, manures and lime on multiple cropped lands. Indian J. Agrl. Sci., 51: 108-111.

Prasad, B., J.Prasad and R.Prasad. 1995. Nutrient management for sustainable rice and wheat production in calcareous soils amended with green manures, organic manures and zinc. Fert. News, 40(2): 39- 45.

Prasad, F.M., K.Singh, V.M.Prasad, M.M.Varma and A.A.Jilani. 2002. Effect of pyrite on the growth and yield of palmorosa under sodic environment. Ann. Pl. Soil Res., 4(2): 349-350.

Prasad, P.V.V., V.Sathyanarayana, M.V.Potdar and P.Q.Craufurd. 2000. On-farm diagnosis and management of iron chlorosis in groundnut. J. Plant Nutr., 23: 1471- 1483.

Prasad, P.V.V., V.Satyanarayana, V.K.K.Murthy and K.J.Boote. 2002. Maximizing yields in rice- groundnut cropping sequence through integrated nutrient management. Field Crops Res., 75: 9-21.

Prasad, R.A. and R.A. Singhania. 1989. Effect of different types of enriched organic manures and time of incubation on soil properties. J. Indian Soc. Soil Sci., 37: 319-322.

Prasad,B., A.P.Singh and N.K.Sinha. 1996. Effect of poultry manures a source of zinc, iron and as a complexing agent on zinc and iron availability and crop yields in calcareous soils. J. Indian Soc. Soil Sci., 44: 519- 521.

Prasert, S. 1997. Application of rice straw compost to lowland rice and its effects on root morphology in Tai paddy fields. In: Proc. IVth JSSRR symposium, pp.32-33.

Pratap Singh , M. K. Singh1 , Sachin Tyagi and Shashank Shekhar Singh.(2018). Effect of Integrated Nutrient Management on Growth and Yield of Rice (Oryza sativa L.) Nripendra, Int.J.Curr.Microbiol.App.Sci (2018) Special Issue-7: 3671-3681

Pratap Bhattacharyya and Datta, S.C. 2005. Effect of organic acids and silicate on phosphorous concentration, buffering capacity and release threshold level to soil. J. Indian Soc. Soil Sci., 53(4): 577- 583.

Pratt, P.F. 1965. In: Methods of analysis (ed.) C.A.Black, American Society of Agronomy Inc., Pub., Madison. Wisconsin. pp.1019- 1021.

Qadar, A. 1995. Requirement of rice crop for phosphorus and potassium at varying sodicity levels. J. Plant Nutr., 18: 2291- 2303.

Rajanirani. 2001. Effect of integration of organics with fertilizer nitrogen on rice and nitrogen uptake. Fertil. News, 46(9): 63- 65.

Rajendra Bhanwaria,Bikarma Singh,Carmelo Maria Musarella. 2022 . Effect of Organic Manure and Moisture Regimes on Soil Physiochemical Properties, Microbial Biomass Plants (Basel). Mar; 11(6): 722.

Rajni Rani and O.P.Srivastava. 2001. Effect of integration of organics with fertilizer N on rice and N uptake. Fertil. News, 46(9): 63-65.

Rama Lakshmi,Ch.S., Rao,P.C., Sreelatha, T., Madahvi, M., Padmaja, G., Rao, P.V. and A. Sireesha.(2012). Nitrogen Use Efficiency and Production Efficiency of Rice Under Rice-Pulse Cropping System with Integrated Nutrient Management Journal of Rice Research.5(1&2).

Ramakrishnan,S., S.Rajagopalan, M.Ramaswamy and A.Ithyarajan. 1992. Effect of Azospirillum on the grain yield of rice. Oryza, 29(2): 161- 162.

Ramaswami, P.P. 1981. Efficient utilization of coir waste. Tamilnadu Agric. Univ. Newsletter, 11- 14.

Rajput, R., P. Pokhriya, P. Panwar, A. Arunachalam and K. Arunachalam. 2019. Soil nutrients, microbial biomass and crop response to organic amendments in rice cropping system in the Shiwaliks of Indian Himalayas Int. J. Recycl. Org. Waste Agric., 8(1): 73-85.

Rao, A.V. and J.C.Tarafdar. 1992. Seasonal changes in available phosphorus and different enzyme activities in arid soils. Annals of Arid Zone, 31: 185-189.

Rao, D.L.N. and H.Pathak. 1996. Ameliorative influence of organic matter on the biological activity of salt affected soils. Arid Soil Res. Rehabilitation, 10: 311-319.

Rathore, A.L., A.R.Pal and K.K.Sahu. 1998. Tillage and mulching effects on water use, root growth and yield of rainfed muster and chickpea grown after lowland rice. J. Sci. Food Agric., 78: 149- 161.

Rathore, R.S, Pyare Lal, P.C.Pandey and P.S.Bisht. 2004. Effect of long term application of fertilizers and farmyard manure on soil fertility and rice productivity under rice-wheat system. Oryza, 41(1&2): 45- 47.

Rathod DD, Rathod PH, Patel KP, Patel KC (2013). Integrated use of organic and inorganic inputs in wheat – fodder maize cropping sequence to improve crop yields and soil properties. Archives of Agronomy and Soil Science 59 (11): 1439- 1455

Ravichandran, V.K., N.Devarani and T.N.Balasubramaniyan. 2003. Dissolution of low-grade rock phosphate in the soil amended with organic manure and Phosphobacteria. J. Ecobiol., 15(1): 55- 59.

Ravindra Kumar, Samir Pal and N.B.Singh. 2006. Yield, uptake and soil fertility in rice-wheat cropping system under integrated nutrient management. Ann. Pl. Soil Res., 8(1): 39- 41.

Reddy, N.S., Mohd Ikramullah, N.V.Ramaiah and C.S.Ramakrishna. 1987. Use of Crassipes and Ipomoea reptans weeds as green manure for rice. Oryza, 24: 79- 86.

Reddy, P.V.R.M. and C.K.Raja Reddy. 2001. Nutrient uptake pattern of sunflower as influenced by integrated nutrient management. The Andhra Agric. J., 48(3&4): 216-222.

Richmond, P.A. 1991. Occurrence and functions of native cellulase. In: Biosynthesis and biodegradation of cellulose. C.H.Haigler and P.J.Weimer. (eds.), Pp. 5- 23, Dekker, New York.

Rivarola, V., S.Castro, G.Mori, E.Jofre, A.Fabra, R.Garnica and H.Balegno. 1998. Response of Azospirillum brasilense Cd to sodium chloride stress. Antonie Leeuwenhoek, 73: 255- 261.

Ron Vaz, M.D., A.C.Edwards, C.A.Shand and M.S.Cresser. 1993. Phosphorus fractions in soil solution: Influence of soil acidity and fertilizer additions. Plant and Soil, 148: 175- 183.

Roy, B. and J.N. Jha. 1987. Effect of treating single super phosphate with cow dung. IRRN, 12 (2): 43.

Rupa, T.R. and I.M.Sukhla. 1998. Adsorption and desorption of zinc and copper in soils of Rayalaseema regions of Andhra Pradesh. J. Indian Soc. Soil Sci., 46(2): 207-211.

Ryden, J.C. and J.K.Syers. 1975. Rationalization of ionic strength and cation effects on phosphorus sorption by soils. J. Soil Sci., 26: 395-406.

Satyanarayana,, V., P. V. Vara Prasad , V. R.K. Murthy and K. J. Boote.2006. Influence of integrated use of Farmyard manure and inorganic fertilizer on yield and yield component of irrigated low land rice. J.Plant Nutrition.25(10).2081-2090

Sabina Ahmad., A. Basumatary,K.N.Das, B.K.Medhi and A.K.Srivastava. 2014.Effect of Integrated Nutrent Management on yield, nutrient soil fertility in autumn rice in an Inceptisol of Assam.Annals of Plant and Soil Research 16 (3): 192-197 (2014)

Saddler, J.N. 1986. Factors limiting the efficiency of cellulase enzymes. Microbiol. Sci., 3: 84-87.

Sahu, S.K., G.N.Mittra and S.C.Pani. 1994. Effect of zinc sources on uptake of Zn and other micronutrients by rice on vertisol. J. Indian Soc. Soil Sci., 42(3): 487- 489.

Sai Ram,M., Tanmoy Shankar, Sagar Maitra, Rahul Adhikary and G.V.V.S.N. Swamy. 2020. Productivity, nutrient uptake and nutrient use efficiency of summer rice (Oryza sativa) as influenced by integrated nutrient management practices. Crop Res. 55 (3 & 4) : 65-72 .

Sakal, R., B.P.Singh, R.B.Sinha, S.P.Singh and S.M.Jha. 1986. Availability of Zn, Fe, Cu and Mn in sub-himalayas hill and forest soils as influenced by certain soil characteristics. J. Indian Soc. Soil Sci., 34: 191-193.

Sarkar, M.I.U., M.N. Isalm, A. Jahan, A. Islam and J.C. Biswas. 2017. Rice straw as a source of potassium for wetland rice cultivation. Geol. Ecol. Landscapes, 1(3):184-189.

Saleque, M.A., M.J. Abedin, N.I. Bhuiyan, S.K.Zaman and G.M.Panaullah. 2004. Long term effects of inorganic and organic fertilizer sources on yield and nutrient accumulation of lowland rice. Field Crops Res., 86: 53- 65.

Sankavi,R.J., S. Avudaithai , S. Somasundaram and T. Sherene Jennita Rajammal.2020. Effect of Zinc and Iron Enriched Organic Manures on Growth and Yield of Rice under Sodic Soil International Journal of Current Microbiology and Applied Sciences.9 (2):2670-2674.

Santhi, R., S.Jayaraman, K.Mayalagu and A.Gopalswamy. 1991. Efficacy of composted coirpith with different levels of inorganic N on nutrient availability and uptake by rice. In: Proc. of the seminar on utilization of coirpith in agriculture. TNAU, Coimbatore, India. pp. 27- 32.

Santhi. R., G.Selvakumari and R.Perumal. 1998. Targeted yield concept in a rice- rice-residual pulse cropping sequence under IPNSS in Typic ustochrepts of lower Bhavani project area of Tamilnadu. Madras Agric. J., 85(5-6): 230- 235.

Santosh Yadav, BL Yadav, Ajit Kumar Meena, Amit Doodhwal, Poonam Kumari Yadav, Prakash Chand Gurjar, Sunita Choudhary and Saroj Choudhary.2022. Effect of Zn and Fe enriched FYM on soil properties and yield attributes of cowpea [Vigna ungiculata (L.) Wılczek] under sodic water irrigation.The Pharma Innovation Journal.11(2): 380-386.

Sarapatka, B. and M.Krskova. 1997. Interactions between phosphatase activities and soil characteristics at some locations in the Czech Republic. Plant Production, 43(9): 415-419.

Saravanan, A., P.Malarvizhi and K.M.Ramanathan. 1991. Comparative efficiency of coirpith, pressmud and green leaves for improving the rice yield in coastal problem soils of Tanjore district. In: proc. Utilization of coirpith in Agriculture. TNAU, Coimbatore, pp: 143- 150.

Satheesh, N. and N.Balasubramaniyan. 2003. Effect of organic manure on yield and nutrient uptake under rice- rice cropping system. Madras Agric. J., 90(1 - 3): 41- 46.

Savithiri, P., Rani Perumal and R.Nagarajan. 1999. Soil and crop management technologies for enhancing rice production under micronutrient constraints. In: Response and management of nutrients in rice systems. Nut.Cyc. Agro Eco Syst. 53(1): 83- 92.

Savithri, P., S.Subbiah, R.Nagarajan, S.Mani and A.Gopalswamy. 1991. Effect of composted coirpith and gypsum application on soil properties and yield of rice in sodic soil. In: Proc. Utilization of coirpith in agriculture, TNAU, Coimbatore, India, pp. 129- 135.

Scollenberger, C.J. and F.R. Dreibelbis. 1930. Analytical methods in base exchange reaction investigation in soils. Soil Sci., 30: 161- 191.

Selvi, D. and D.Augustine Selvaseelan. 1991. Relative efficiency of composted coirpith as organic manure for rice. In: Proc. Utilization of coirpith in agriculture, TNAU, Coimbatore, India, pp. 73- 82.

Selvi,D., P.Santhy and M. Dhakshinamoorthy. 2005. Effect of inorganic alone and in combination with farmyard manure on physical properties and productivity of vertic haplusteps under long-term fertilization. J. Indian Soc. Soil Sci., 53(3); 302- 307.

Selvi,D., S.Mahimai Raja, P.Santhy and B.Rajkannan. 2003a. Influence of integrated nutrient management on crop yield, P uptake and available P status of soil in a rice based cropping system. Oryza, 40(3&4): 65- 69.

Selvi,D., P.Subramaniyan, K.R.Latha and R.Kalpana. 2003b. Organics in the improvement of nutrient use efficiencies in white ponni. Madras Agric. J., 90(10- 12): 611- 614.

Sen, H.S. and B.K.Bandyopadhyay. 1986. Effect of placement, urease inhibitors and acidic amendments on ammonia volatilization from wetland rice field fertilized with urea. J. Indian Soc. Coastal Agric. Res., 34: 514- 519.

Sensabaugh, R.L., E.F.Benfield and A.E.Linkins. 1981. Cellulase activity associated with the decomposition of leaf litter in a woodland streams. Oikos, 3: 84-87.

Seshadri, S., R.Muthukumarasamy, E. Laxminarashiman and S. Ignacimuthu. 2000. Solubilization of inorganic phosphates by Azospirillum halopraferans. Curr. Sci., 79: 565- 567.

Shaaban, M.; Abid, M.; Abou, S. Amelioration of salt affected soils in rice paddy system by application of organic and inorganic amendments. Plant Soil Environ. 2013, 59, 227–233.

Shambhavi S, Kumar R, Sharma SP, Verma G, Kumar S, Sharma K, Sharma RP. 2018. Effect of 36 Years of Continuous Cropping and Fertilization on Productivity, Micro and Secondary Nutrient Status and Uptake by Maize-wheat Cropping System in Western Himalayas. International Journal of Bio resource and stress Management, DOI: TTPS://DOI.ORG/10.23910/IJBSM/ 2018.9.2.1854b 197-202.

Shanmugam, P.M and Veeraputran. 2000. Effect of organic manure, biofertilizers, inorganic nitrogen and zinc on the yield and growth of rabi rice. Madras Agric. J., 87(103): 90-92.

Sharma, A.R. and B.N. Mittra. 1991. Effect of combinations of organic materials and nitrogen fertilizers on growth, yield and nitrogen uptake of rice. J. Agric. Sci. Camb., III: 605- 608.

Sharma, A.R. and K.C.Das. 1994. Effect of green manuring with dhaincha (Sesbania aculeata) on growth and yield of direct sown and transplanted rice under intermediate deep water conditions (0- 50 cm). J. Agric. Sci., 122(3): 359- 364.

Sharma, M.P., S.V.Balu and D.K.Gupta. 2001. Soil fertility and productivity of rice- wheat cropping system in an inceptisol as influenced by integrated nutrient management. Indian J. Agric. Sci., 71: 82- 86.

Sharma, S.N. and R.Prasad. 1999. Effects of sesbania green manuring and mungbean residue incorporation on productivity and nitrogen uptake of a rice wheat cropping system. Bioresour. Technol., 67: 171- 175.

Sharma, S.N., R.Prasad and R.K.Singh. 2000. Influence of summer legumes in rice- wheat system on soil fertility. Indian J. Agric. Sci., 70: 357- 359.

Sharma U, Subehia SK (2014). Effect of long term integrated nutrient Management on rice (Oryza sativa L.)-wheat (Triticum aestivum L.) productivity and soil properties in north-western Himalaya. Journal of Indian Society of Soil Science 62(3): 248-254.

Sheeba, S. and S.Chellamuthu. 1999. Long term influence of organic and inorganic fertilization in the micronutrient status of Inceptisols. J. Indian Soc. Soil Sci., 47: 803- 804.

Shivakant, P. and R.Rajkumar. 1992. Effect of gypsum, pressmud and farmyard manure on soil properties and yield of rice. Indian J. Agric. Sci., 62(3): 489- 492.

Shormy,S.A.S., M.A.H. Chowdhury, B.K. Saha and M.A. Haque . 2013. Effects of different sources of organic materials on nutrient contents and their uptake by T. aman rice. J. Agrofor. Environ. 7 (1): 37-40,

Shyamal Kheroar., Bikas Chandra Patra., Debjani Halder., Swapan Kumar Barman and Kausik Mandal. (2018). Comparative Efficacy of Inorganic and BioFertilizers on Growth and Yield of Rainfed Winter Rice (Oryza sativa L.). Current Journal of Applied Science and Technology.26(2):1-3.

Sidhu GS, Sharma B.D. 2010. Diethylenetriamine pentaacetic acid-extractable micronutrients status in soil under a rice-wheat system and their relationship with soil properties in different agroclimatic zones of Indo-gangetic plains of India. Communication of Soil Science and Plant Analysis, 41, 29-51

Singaravel, R. and V. Prasath. 2004. Effect of bio-resources on the nutrient availability and microbial activity in degraded coastal soil. J. Indian Soc. Coastal Agric. Res., 22(1&2): 61- 64.

Singaravel, R., V.Imayavaramban, K.Thanunathan and K.Johnson. 2001. Effect of gypsum and organic residues on the amelioration of coastal saline soil. Ecoprint, 8(1): 103-105.

Singh, G., Singh ,O.P., Singh, R.G., Mehta,R.K, Kumar,V. and Singh,R.P. 2006. Effect of integrated nutrient management on yield and nutrient uptake of rice (Oryza sativa)– wheat (Triticum aestivum) cropping system in lowlands of eastern Uttar Pradesh.Indian Journal of Agronomy. Year : 2006, : 51 (2): 85-88..

Singh Parihar, A.K., J.Singh, B.S.Rana & Govind Swarup. 2003. Effect of nitrogen sources on yield and N- uptake by rice in a sodic soil. Ann. Pl. Soil Res., 5(2): 232- 233.

Singh, F., R.Kumar and S.Pal. 2003. Integrated nutrient management in rice- wheat cropping system for sustainable productivity. Prog. Agric., 3: 115- 116.

Singh, K.D., R.N.S.Verma and S.S.Lohia. 1980. Effect of continuous application of FYM and chemical fertilizer on some soil properties. J. Indian Soc. Soil Sci., 28: 170- 172.

Singh, K.N., A.Swarup, D.P.Sharma and K.V.G.K.Rao. 1990. Influence of subsurface drainage and phosphorus application on soil salinity and uptake of phosphorus by wheat in saline soil. J. Indian Soc. Soil Sci., 38: 511-515.

Singh, K.N., A.Swarup, D.P.Sharma and K.V.G.K.Rao. 1992. Effect of drain spacing and phosphors levels on yield, chemical composition and uptake of nutrients by Indian mustard (Brassica juncea) Experiment. Agric., 28: 135-142.

Singh, S. and R.N.Roy. 1975. Effect of salinity, alkalinity, phosphate and age of plants on mineralization of nitrogen from Sesbania aculeata and Melilotus alba after their incorporation in soil. J. Indian Soc. Soil Sci., 23: 122.

Singh, S.K. and B.C.Ghosh. 1999. Integrated nutrient management in jute- rice system under rainfed lowlands. Indian J. Agric. Sci., 69: 300- 301.

Singh, A., A.P. Singh, S.K. Singh, S. Rai and D. Kumar. 2016. Impact of addition of biochar along with PGPR on rice yield, availability of nutrients and their uptake in alluvial soil. J. Pure Appl. Microbiol., 10(3): 2181-2188.

Singhal, S.K., V.K. Sharma and S. Teekam. 2012. Integrated effect of enriched compost and fertilizer on yield and uptake of micronutrient by maize (Zea mays L.). Agric. Sci. Digest, 32(1): 43-47.

Sonune, B.A., K.B.Tayade, V.V.Gabhane and R.B.Puranik. 2003. Long term effect of manuring and fertilization on fertility and crop productivity of vertisols under sorghum- wheat sequence. Crop Res., 25(3): 460- 467.

Speir, T.W. and D.J.Rose. 1981. Studies on a climosequence of soils in tussock grasslands. 24. Enzyme activities of tussock litter exposed around the base of tussock plants. New Zealand J. Sci., 24: 145-151.

Speir, T.W., E.A.Pansier and A.Cairns. 1980. A comparison of sulphatase, urease and protease activities in planted and fallow soils. Soil Biol. Biochem., 12: 281-191.

Sreedevi, B. and G.S.Thangamuthu. 1991. Studies on NPK uptake by rice under integrated nitrogen management. Oryza, 28: 255-256.

Sridevi, S., J.C.Katyal, K.Srinivas and K.L.Sharma. 2003. Carbon mineralization and microbial biomass dynamics in soil amended with plant residues and residue fractions. J. Indian Soc. Soil Sci., 51(2): 133- 139.

Srikanth, K., K.A.Srinivasamurthy, R.Siddaramappa and V.R.Ramakrishna Parama. 2000. Direct and residual effect of enriched compost, FYM, vermicompost and fertilizers on properties of an alfisol. J. Indian Soc. Soil Sci., 48: 496- 499.

Srinivasa Rao, CH., A.Subba Rao and S.K.Bansal. 1996. Potassium supply and management through non- conventional sources for optimizing K nutrition and crop production. Fertil. News, 41(7): 33- 40.

Sriramachandrasekaran, M.V. and V. Vaiyapuri. 2002. Effect of green manure, fertilizer nitrogen and cytozome on enzyme activities in different soils. J. Ecobiol., 14(4): 247-250.

Sriramachandrasekaran. M.V., R.Bhuvaneswari and M. Ravichandran. 2004. Integrated use of organics and sulphur on the rice yield and sustainable soil health in sulphur deficient soil. Plant Archives 4(2): 281- 286.

Sriramachandrasekharan, M V. 2001. Effect of organic manures on the nutrient uptake, yield and nutrient use efficiency in lowland rice. Journal of Ecobiology 13(2): 143-147.

Stanford, S. and L.English. 1949. Use of flame photometer in rapid soil test for K and Ca. Agron. J., 41: 446- 447.

Subba Rao, I.V., M.V.Shantaram and V.Adinarayana. 1994. Soil constraints and strategies for efficient land management in coastal areas. J. Indian Soc. Coastal Agric. Res., 12(1&2): 23- 35.

Subba Rao, N.S. 1995. Biological nitrogen fixation: Potentialities, prospects and limitations. In: Biological nitrogen fixation: Recent developments. N.S.Subba Rao (ed.). Oxford & IBH publishers, New Delhi. Pp: 1-9.

Subba Rao, N.S. 1999. Soil Microbiology. IVth edition. Soil microorganisms and plant growth. Oxford & IBH Publishing Co. Pvt. Ltd., New Delhi.

Subbaiah, G. and A.L.Asija. 1956. A rapid procedure for estimation of available nitrogen in soils. Curr. Sci., 125: 259- 260.

Subehia, S.K. and R.S.Minhas. 1993. Phosphorus availability from Udaipur Rock Phosphate as influenced by different organic amendments. J. Indian Soc. Soil Sci., 41(1): 96- 99.

Subramani,S. 1986. Studies on the decomposition of Azolla in rice soil. M.Sc.(Agri.) thesis, Tamilnadu Agricultural University, Coimbatore, Tamilnadu, India.

Subramani,S. and S.Kannaiyan. 1989. Effect of water fern Azolla on enzyme activity in soil under rice cultivation. Mikrobiologiya, 58(1): 118- 121.

Subramaniyan, S. and A. Gopalaswamy. 1991. Effect of moisture, organic matter, phosphate and silicate on availability of silicon and phosphorus in rice soils. J. Indian Soc. Soil Sci., 39: 99- 103.

Sud, K.C., J.S.Grewal and S.P.Trehan. 1990. Effect of farmyard manure and nitrogen on potato (Solanum tuberosum) production and phosphorus and potassium availability in hill soils of Shimla. Indian J. Agric. Sci., 60 (8): 529- 532.

Sultan Singh, Satyavir Singh, O.P.Dhankar and Abha Tikoo. 2005. Importance of gypsum in crop production under sodic water use in light textured soils of Haryana. J. Indian Soc. Soil Sci., 53(2): 249- 253.

Sumen Metha, Anil Kumar and N.P.Thakur. 2004. Residual effects of integrated nutrient management on growth and yield of succeeding rice (Oryza sativa L.) crop in wheat (Triticum aestivum) – rice-cropping system. Oryza, 41(3&4): 105- 107.

Sundara Rao, W.V.B. and M.K.Sinha. 1963. Phosphate dissolving organisms in the soil and rhizosphere. Indian J. Agric. Sci., 33: 272-278.

Sundaravarathan, S. and S. Kanniayan. 1998. Decomposition of Azolla microphylla and Sesbania rostrata with fertilizer N and neem cake under submerged soil conditions. Indian J. Microbiol., 38: 89- 93.

Surendra Rao, S. and M.Sitaramayya. 1997. Changes in soil total and available nitrogen under integrated nutrient management of rice. J. Indian Soc. Soil Sci., 45(3): 42- 43.

Suresh, K.K., K.Ramamoorthy and J. Sunder Singh Raja Pandiyan. 2001. Integrated nutrient management to economise phosphorus and potassium in rainfed rice. Madras Agric. J., 88(10- 12): 732- 733.

Swarup, A. 1987. Effect of presubmergence and green manuring (Sesbania aculeata) on nutrition and yield of wetland rice on a sodic soil. Biol. Fertil. Soils, 5(3): 203- 208.

Swarup,A. and N.P.S.Yaduvanshi. 2000. Effects of integrated nutrient management on soil properties and yield of rice in alkali soils. J. Indian Soc. Soil Sci., 48: 279- 282.

Syed Nazeer Peeran and U.S. Sree Ramulu. 1995. Influence of urea, organic manures, Azospirillum and growth regulators on the yield of rice. Oryza 32: 28- 32.

Szalai T, Lehoczky E, Nyarai H, Hollo S, Ksatho P(2002). The available micro-element content of soil in a long-term nutrient supply experiment. Communication in Soil Science and Plant Analysis33: 3251-3260

Szegi, J. 1988. Cellulose decomposition and soil fertility. Akademiai Kiado, Budapest.

Takeda A, Hirafumi Tsukada, Masami Nanzyo, Yuichi Takaku, Toyokazu Uemura, Shunichi Hisamatsu (2005). Effect of Long Term Fertilizer Application on the Concentration and Solubility of Major and Trace Elements in a Cultivated Andisol. Soil Science and Plant Nutrition 51(2):251—260.

Tabatabai, M.A and J.M. Bremner.1972. Assay of urease activity in soil. Soil Biol. Biochem., 4: 479-487.

Tabatabai, M.A. and J.M.Bremner. 1969. Use of p-nitrophenyl phosphate for assay of soil phosphatase activity. Soil Biol. Biochem., 1: 301-307.

Tajeda, M.C.,Garcia,J.Gonzalez, MJSB Hernandez. 2006. Use of organic amendments as a strategy for saline soil remediation: influence on the physical, chemical and biological properties of soil. Soil Biol. and Biochemistry. 38(6):1413-1421.

Talashilkar, S.C. and A.S.Chavan. 1997. Integrated management of green leaf manure and urea-nitrogen for wetland rice in Konkan region. Ann. Agric. Res., 18(1): 124- 126.

Tan, K.H. 1986. Degradation of soil minerals by organic acids. In: Huang, P.M, Schnitzer, M. (eds.) Interactions of soil minerals with natural organics and microbes. Soil Sci. Soc. Am., Madison, Wisconsin, pp.1-25.

Tarafdar, J.C. and N.Claassen. 1988. Organic phosphorus compounds as a phosphorus source for higher plants through an activity of phosphatase produced by plant roots and microorganisms. Biol. Fertil. Soils, 5: 308-312.

Tarafdar,J.C., V.A. Rao and K.Bala. 1988. Production of phosphatases by fungi isolated from desert soils. Fol. Microbiol., 33: 453-457.

Tester, C.F. 1989. Tall fescue growth in green house growth chamber and field plots amended with sewage sludge compost and fertilizers. Soil Sci., 148: 452-458.

Thakuria, K., B.Borgahain and K.K.Sharma. 1991. Effect of organic and inorganic sources of nitrogen with and without phosphate on fiber yield of white jute (Corchorus capsulanes). Indian J. Agric. Sci., 61(1): 49-50.

Thangaraju, M. and S.Kanniyan. 1990. Influence of Azolla, Sesbania and urea super granule on rice yield and nitrogen uptake. IRRN, 15(1): 24.

Thanikachalam,A., K.Rajkannu and S. Kanniyan. 1984. Effect of neem cake, carbofuron and Azolla application on phosphatase activity in soil. In: 25th Ann. Conf. Assoc. Microbiologists of India. Pant Nagar, p. 7.

Thompson, J.F., I.K.Smith and J.T.Madison. 1966. Sulphur metabolism in plants. In: Tabatabai, M.A. (ed.): Sulfur in agriculture. Agron. Mohogr., 27, ASA, Madison, WI, pp.295-322.

Thompson, M.L., Heinlin Zhang, Masoud Kazemi and J.A. Sander. 1989. Contribution of organic matter to cation exchange capacity and specific surface area of fractionated soil materials. Soil Sci., 148(4): 250-257.

Thuler, D.S., Floh, E.I.S., W. Handro and H.R. Barbosa. 2003. Plant growth regulators and amino acids released by Azospirillum sp. in chemically defined media. Lett. Appl. Microbiol., 37: 174-178.

Tiarks, A.E., A.P.Mazurak and Z. Chesnin. 1974. Physical and chemical properties of soil associated with heavy application of manure from cattle feedlots. Soil Sci. Soc. Am. Proc., 38: 826-830.

Tilahun Tadesse, Nigussie Dechassa, Wondimu Bayu, Setegn Gebeyehu 2013. Effects of Farmyard Manure and Inorganic Fertilizer Application on Soil Physico-Chemical Properties and Nutrient Balance in Rain-Fed Lowland Rice Ecosystem. American Journal of Plant Sciences. Vol.4 No.2 (2013).

Tiwari, K.N. and D.N.Sharma. 1989. Soil salinity research. All India coordinated project on management of salt affected soils and use or saline water in agriculture, CSAUAT, Khanpur, pp.89.

Tiwari, K.N., A.N.Pathak and Hariram. 1980. Green manuring in combination with fertilizer N on rice under double cropping system in an alluvial soil. J. Indian Soc. Soil Sci., 28: 162-169.

Tiwari, V.N., Harisingh and R.M.Upodhyay. 2001. Effect of biocides, organic manure and blue green algae on yield attributing characteristics of rice and soil productivity under sodic soil condition. J. Indian Soc. Soil Sci., 49: 332-336.

Toth, S.J. and A.L. Prince. 1949. Estimation of cation exchange capacity and exchangeable calcium, potassium and sodium contents of soil by Flame photometer. Soil Sci., 67: 439-445.

Tripathi,B.R. 1998. Managing sodic soils for sustained crop production in Indo- Gangetic plains. J. Indian Soc. Soil Sci., 46(4): 543-550.

Uma Shanker Ram, VK Srivastava, A Hemantaranjan, A Sen, RK Singh, JS Bohra and U Shukla. 2013.Effect of Zn, Fe and FYM application on growth, yield and nutrient content of rice. Oryza. 50(4):351-357.

Vanaja, M. and Sreenivasa Raju. 2003. Integrated nutrient management in rice- sunflower cropping system in alfisols. J. Soils and Crops 13 (2): 211-218.

Vance, W.H.; Tisdall, J.M.; McKenzie, B.M. Residual effects of surface applications of organic matter and calcium salts on the subsoil of a red-brown earth. Aust. J. Exp. Agric. 1998, 38, 595–600.

Verma A, Nepalia V, Kanthaliya PC.(2005) Effect of continuous cropping and fertilization on crop yields and nutrient status of a Typic Haplustept. Journal of the Indian Society of Soil Science.53, 365-368

Velmurugan, A. and T.P. Swarnam. 2021. Effect of enriched organic manures and traditional practices on soil carbon dynamics and yield of maize-rice cropping system in coastal lowlands. J. Pharmacogn. Phytochem., 10(1): 273-278.

Vijayaprabhakar, A., S.N. Durairaj and J.V. Raj. 2017. Effect of rice straw management options on soil available macro and micro nutrients in succeeding rice field. Int. J. Chem. Stud., 5(4): 410-413.

Vig, A.C., Didar Singh, Milap Chand and G.S.Saroa. 1997. Release of phosphorus from Sesbania aculeata. J. Indian Soc. Soil Sci., 45(3): 449-455.

Vimal, O.P. and S. Talashikar. 1983. Recycling of wastes in agriculture: looking ahed to 2000 AD. J. Scientific Industr. Res., 42: 115-127.

Vinothkumar, V., B.C.Ghosh and B.Ravi. 1998. Recycling of crop wastes and green manure for sustained soil fertility under rice based cropping system. In: Extended Summaries of First International Agronomy Congress, 23-27 November, 1998, New Delhi, India, Pp. 434-435.

Visser,S. and Parkinson, D. 1992. Soil biological criteria as indicators of soil quality: soil microorganisms. Am. J. Altern. Agric., 7: 33-37.

Von Unexlkull, H.R. and E. Mutert. 1992. Principles of balanced fertilization. In: Proceedings of Regional FADINAP Seminar on Fertilization and Environment, Chiang mai, Thailand, 7-11 September 1992.

Walkley, A. and C.A.Black. 1934. An examination of the degtijareft method of determination of organic matter and proposed modification of the chromic acid titration method. Soil Sci., 37: 29-38.

Wan, J.H.C. and M.H. Wong. 2004. Effects of earthworm activity and P- solubilizing bacteria on P availability in soil. J. Plant Nutr. Soil Sci., 167: 209-213.

Wang, M.C. and C.H.Yang. 2003. Type of fertilizer applied to a paddy upland rotation affects selected soil quality attributes. Geoderma, 114: 93- 108.

Wang, L.; Sun, X.; Li, S.; Zhang, T.; Zhang, W.; Zhai, P. Application of organic amendments to a coastal saline soil in North China: Effects on soil physical and chemical properties and tree growth. PLoS ONE 2014, 9, e89185. [Google Scholar] [CrossRef].

Wani, S.P. 1990. Inoculation with associative nitrogen bacteria: Role in cereal grain production and improvement. Indian J. Microbil., 30: 363- 393.

Wen Qixiao and Yu Tianren. 1988. Effect of green manure on physico-chemical properties of irrigated rice soils. In: Proc. of Symposium on sustainable agriculture. Green manure in rice farming. IRRI, Philippines, pp. 275- 287.

Wheaton, A.D.; McKenzie, B.M.; Tisdall, J.M. Management of a sodic soil for wine grape production. Aust. J. Exp. Agric. 2002, 42, 333–339.

Yadav, J.S.P., A.K.Bandyopadhyay and B.K.Bandyopadhyay. 1983. Extent of coastal saline soils in India. J. Indian Soc. Coastal Agric. Res., 1: 1-16.

Yadav, R.L. 2004. Enhancing efficiency of fertilizer N use in rice- wheat systems of Indo-Gangetic plains by intercropping Sesbania aculeata in direct seeded upland rice for green manuring. Bioresour. Technol., 93: 213- 215.

Yadav, R.L., B.S.Dwivedi, Kamta Prasad, O.K.Tomar, N.J.Shurpali and P.S.Pandey. 2000. Yield trends and changes in soil organic- C and available NPK in a long-term rice-wheat system under integrated use of manures and fertilizers. Field Crop Res., 68: 219- 246.

Yadav. J.S.P. 1980. Efficiency of fertilizers used in saline and alkali soils for crop production. Fertil. News, 25(a): 19- 27.

Yadav, L. and Meena .N. (2014) Performance of aromatic rice (Oriza sitiva) genotype as influenced by integrated nitrogen management. Indian Journal of Agronomy 59(2): 51-255

Yadav, D.S, Kumar,A. (2009). Long-term effect of nutrient management on soil health and productivity of rice (Oryza Sativa)-wheat (Triticum Aestivum) system. Indian Journal of Agronomy 54(1): 15-23.

Yaduvanshi, N.P.S. 2001. Effect of five years of rice- wheat cropping and NPK fertilizer use with and without organic and green manures on soil properties and crop yields in a reclaimed sodic soil. J. Indian Soc. Soil Sci., 49(4): 714- 719.

Yaduvanshi, N.P.S.,D.R.Sharma and A. Swarup.2013. J.Indian Soc. Soil Sci.61.188-194.

Yadvinder Singh, Bijai Singh and C.S. Khind. 1992. Nutrient transformations in soils amended with green manures. Adv. Soil Sci., 20: 237- 307.

Yadvinder Singh, Bijay Singh, M.S.Maskina and O.P.Meelu. 1988. Effect of organic manures, crop residues and green manure (Sesbania aculeata) on nitrogen and phosphorus transformations in a sandy loam soils at field capacity and waterlogged conditions. Biol. Fertil. Soils, 6: 183- 187.

Yadvinder Singh, R.P.S.Pannu, Bijay Singh and C.S.Khind. 2005. Leaching of potassium from organic manures, crop residues and inorganic fertilizer in two soils under flooded and upland moisture regimes. J. Indian Soc. Soil Sci., 53(2): 207- 213.

Yang, C.M., L.Z.Yang and T.M.Yan. 2002. Effects of different nutrient and water regimes on paddy soil eco-environment. Rural Eco- Environ., 18: 11-15.

Yoav Bashan, Gina Holguin and Luz E. de- Bashan. 2004. Azospirillum- plant relationships: physiological, molecular, agricultural and environmental advances (1997- 2003). Can. J. Microbiol. 50: 521- 577.

Yogananda, S.B. and V.C.Reddy. 2004. Integrated nutrient management for hybrid rice KRH- 2. Mysore J. Agric Sci., 38(3): 319- 327.

Yoshida, S. 1981. Fundamentals of rice crop science. The International Rice Research Institute, Manila, Philippines.

Yu, T. (ed.). 1985. Physical chemistry of paddy soils. Science Press, Beijing and Springer-Verlag, Berlin, Heidelberg.

Yung- Yu Shu. 2005. Effect of application of different types of organic composts on rice growth under laboratory conditions. Soil Sci. Plant Nutr., 51(3): 443- 449.

Zhao, X., G. Yuan, H. Wang, D. Lu, X. Chen and J. Zhou. 2019. Effects of full straw incorporation on soil fertility and crop yield in rice-wheat rotation for silty clay loamy cropland. Agron., 9: 133.

Zeb, H., A. Hussain, M. Naveed, A. Ditta, S. Ahmad, M.U. Jamshaid, H.T. Ahmad, M.B. Hussain, R. Aziz and M.S. Haider. 2018. Compost enriched with ZnO and Zn-solubilizing bacteria improves yield and Zn-fortification in flooded rice. Ital. J. Agron., 13(4): 310-316.

Zantua, M.I. and J.M. Bremner. 1977. Stability of urease in soils. Soil Biol. Biochem., 9: 135- 140.

Zhou, Y. and A.C.Luo. 1997. Effect of organic manure on phosphorus absorption and root activity of wheat. Plant Nutr. Fertil. Sci., 3: 243-248.

Zhu, H. H., J.S. Wu, D.Y. Huang, Q.H. Zhu, S.L. Liu, Y.R. Su, W.X. Wei, J.K. Syers and Y. Li. 2010. Improving fertility and productivity of a highly-weathered upland soil in subtropical China by incorporating rice straw. Plant Soil, 331: 427-37.